Emerging Technologies, Environment and Research for Sustainable Aquaculture

Edited by Qian Lu
and Mohammad Serajuddin

Published in London, United Kingdom

IntechOpen

Supporting open minds since 2005

Emerging Technologies, Environment and Research for Sustainable Aquaculture
http://dx.doi.org/10.5772/intechopen.82887
Edited by Qian Lu and Mohammad Serajuddin

Contributors
V P Sharma, Hai-Hong Huang, Francois Gagné, Brian Quinn, Joëlle Auclair, Maruf Kasim, Abdul Muis Balubi, Wardha Jalil, Ahmad Mustafa, Rahman Nurdin, Rahmat Sofyan Patadjai, Firuza Mustafa, Bwadi Benjamin Ezekiel, Shao-Wen Hung, Chung-Lun Lu, Shiu-Nan Chen, André Freccia, Maurício Gustavo Coelho Emerenciano, Artur Rombenso Paquito, Jiovani Sergio Bee Tubin, Jorge Parodi, Viviana Chavez, Matias Peredo, Jayasankar Pallipuram, Arun Raj Velraj

First published in London, United Kingdom, 2020 by IntechOpen
IntechOpen is the global imprint of INTECHOPEN LIMITED, registered in England and Wales, registration number: 11086078, 7th floor, 10 Lower Thames Street, London, EC3R 6AF, United Kingdom
Printed in Croatia

British Library Cataloguing-in-Publication Data
A catalogue record for this book is available from the British Library

Additional hard and PDF copies can be obtained from orders@intechopen.com

Emerging Technologies, Environment and Research for Sustainable Aquaculture
Edited by Qian Lu and Mohammad Serajuddin
p. cm.
Print ISBN 978-1-83881-199-0
Online ISBN 978-1-83881-200-3
eBook (PDF) ISBN 978-1-83881-201-0

For EU product safety concerns:
IN TECH d.o.o., Prolaz Marije Krucifikse Kozulić 3, 51000 Rijeka, Croatia, info@intechopen.com or visit our website at intechopen.com.

Meet the editor

Qian Lu received his Ph.D from the University of Minnesota (USA) and then worked at Nanchang University (China). His research interests include green technology for aquaculture and the remediation of eutrophic wastewater. Until now, in the aspects of environmental protection and aquaculture, he has published over 30 peer-reviewed papers as first author or corresponding author. Now he is working on several industrial projects related to sustainable and eco-friendly aquaculture systems.

Dr. M. Serajuddin is presently a Professor in the Department of Zoology, University of Lucknow, U.P., India. He is a Fellow of the Zoological Society of India and Academy of Environmental Biology. He has received several awards of national and international reputes. He has been involved in research and teaching for over 25 years in fish biology, prevention of cancer through fish lipids, and fish biodiversity. Dr. Serajuddin acted as Principal Investigator for major research projects funded by the Government of India. He has supervised 15 Ph.D. students in zoology. He has to his credit more than 80 original research papers in peer-reviewed journals of national and international repute. He has also edited 6 books bearing ISBN numbers.

Contents

Preface

Aquaculture is a promising industry that provides humans with high-quality protein. However, to our knowledge, there are many problems, such as water body eutrophication and antibiotics abuse, which jeopardize the sustainable development of aquaculture. For the purpose of sustainable development, new technologies and policies are highly needed in the field of aquaculture and fisheries. Therefore, to promote the sustainable development of aquaculture, this book focuses on the emerging technologies and research for sustainable aquaculture.

The book is divided in two sections: Technologies and Environment, and Sustainable Aquaculture It is expected that this book can help researchers and technicians in the aquaculture industry get more new ideas and techniques.

Authors from different countries and institutes contributed to the chapters of this book. We would like to highly commend them on their great contributions.

Qian Lu
School of Resources, Environmental and Chemical Engineering,
Nanchang University,
Nanchang, China

Mohammad Serajuddin
Department of Zoology,
University of Lucknow,
U.P., India

Section 1

Technologies and Environment

Chapter 1

Novel Biofloc Technology (BFT) for Ammonia Assimilation and Reuse in Aquaculture In Situ

Hai-Hong Huang

Abstract

Ammonia is one of the most harmful risks for success of fish and shrimp culture. There is no effective solution for harmlessness of ammonia in traditional aquaculture operations except exchanging water, which would bring negative effects on environment, or fixing expensive equipment. Biofloc technology (BFT) that appeared in recent years supplies a novel solution for this issue without exchanging huge water and fixing equipment. This technology could assimilate ammonia almost in real time with many other supplemental benefits. Because of the very high nutritional value for fish and shrimp, bioflocs, the by-product of BFT, could also be reused as a complemented food in situ or a gradient for feedstuff to replace expensive fishmeal or be processed to pellet diet to feed fish and shrimp directly. However, some aspects with regard to the effective use of biofloc as a food source for fish and shrimp, such as high lipid content, productivity, and palatability, need to be further researched in detail.

Keywords: ammonia, assimilation, reuse, biofloc technology, aquaculture

1. Introduction

The world population will exceed 9 billion people by the middle of the twenty-first century, indicating proportionate food should have to provide. Fisheries and aquaculture are the critical important sources against this challenge of food and nutrition [1]. Between 1961 and 2016, the average annual increase in global food fish consumption (3.2%) outpaced population growth (1.6%) and exceeded that of meat from all terrestrial animals combined (2.8%). Total fish production in 2016 reached 171 million tones, of which 88% was directly utilized for human consumption. In per capita terms, food fish consumption grew from 9.0 kg in 1961 to 20.2 kg in 2015, accounting for about 17% of their average per capita intake of animal protein consumed by the global population [1].

Since the late 1980s, the fishery production has been stable without obvious increase. But aquaculture has become more and more important, which production grew faster than other major food production sectors. The contribution of aquaculture to the global production of capture fisheries and aquaculture combined has risen continuously, reaching 46.8% in 2016 and representing 53% of fish production for food uses [1].

However, the development of aquaculture has faced challenges because of lack of land and water source and degradation of environment [1]. Therefore, turning of aquaculture to intensive even high intensive model from extensive or semi-extensive model is a tendency all over the world. Intensive aquaculture utilizes limited land source to culture more fish and shrimp by excessively increasing aquatic animal density with little water exchange or even zero water exchange. However, one of the most harmful risks for success of fish and shrimp in intensive aquaculture system, especially in closed intensive culture system with little water exchange, is the accumulation of ammonia. Unfortunately, there is no effective solution for harmlessness of ammonia in practical operations except exchanging water or fixing some very expensive equipment for water treatment [2].

Biofloc technology (BFT) that appeared in recent years supplies a novel solution for this issue without exchanging huge water or fixing equipment [2]. BFT could assimilate ammonia almost in real time and reuse the by-product as a natural food source in situ in aquaculture water column. In this chapter, problems referred to ammonia in aquaculture (Part 2 and Part 3), principles of ammonia removal (Part 4), main operations of BFT (Part 5), applications of using biofloc produced as a by-product of BFT in aquaculture (Part 6 and Part 7), and some highlighting issues that should be paid attention to or need to be further researched (Part 8) are introduced in brief.

2. Toxicity of ammonia to fish or shrimp

Ammonia is one of the most harmful inorganic nitrogen compounds for fish or shrimp in aquaculture (another is nitrite), whose accumulation in pond water may deteriorate water quality, reduce growth, increase oxygen consumption, alter concentrations of hemolymph protein and free amino acid levels, and even cause high mortality [3]. For example, in water with pH 8.05 and temperature 23°C, the 96 h median lethal concentration (LC_{50}) value of ammonia on *Litopenaeus vannamei* (Pacific white-leg shrimp, the most important cultured crustacean species in the world [1]) juveniles is 35.4 mg/L at salinity of 25‰, indicating that the safety level of ammonia for rearing *L. vannamei* juveniles is only 3.55 mg/L [3]. A research conducted by the author of this chapter revealed that in a closed aquaculture system for *L. vannamei* without water exchange, the ammonia concentration accumulated up to a very high level of 10.81 mg/L, with an average of 6.35 mg/L, leading to a mortality rate as high as 70%.

There are two existent types for ammonia, ionic type (NH_4^+) and free type (NH_3), both of which in general named together as total ammonia nitrogen (TAN). In fact, the toxicity of TAN is mainly from the free NH_3; in water, the 96 h LC_{50} value to *L. vannamei* juveniles is 1.57 mg/L under conditions of pH 8.05, temperature 23°C, and salinity of 25‰, and the safety level is 0.16 mg/L [3]. Van Wyk et al. [4] even recommend a value of ≤0.03 ppm or mg/L for NH_3 when farming this shrimp in recirculating freshwater systems. There is an equilibrium between the two existences of ammonia in water [5]:

$$NH_3 + H^+ \leftrightharpoons NH_4^+ \tag{1}$$

This equilibrium indicates that NH_4^+ and NH_3 exist in water at the same time and their proportions are determined by the pH of the water body so that the toxicity of TAN is highly related to water pH. Actually, the relationship among water pH and the concentrations of NH_4^+ and NH_3 could be descripted with an equation [6] as follows:

$$\frac{[NH_3]}{[NH_3] + [NH_4^+]} = \frac{10^{pH}}{\exp\left(\frac{6344}{273+T}\right) + 10^{pH}} \tag{2}$$

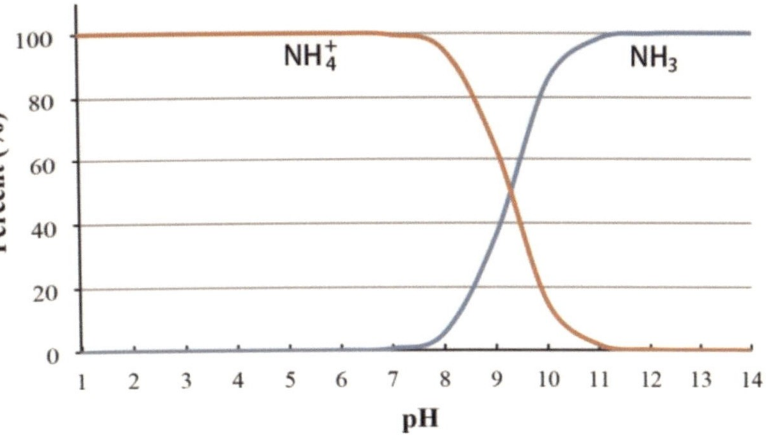

Figure 1.
Percent of NH_4^+ and NH_3 changing with pH under 25°C.

In this equation, *T* means the water temperature. Based on Eq. (1), the proportions of NH_4^+ and NH_3 along pH gradient under a certain temperature (such as 25°C) could be calculated and pictured, respectively (**Figure 1**). From **Figure 1**, it is well known that there is almost no NH_3 existing in water when pH is below 7.0; however, after that, the proportion of NH_3 will exponentially elevate with increasing of pH, as well as the toxicity of TAN.

3. Deriving of ammonia in aquaculture water body

Ammonia in aquaculture water body is mostly produced from artificial feeds fed to fish animals. Estimated about 78% of nitrogen existing in aquaculture water body comes from feedstuff [7]. Artificial formulated feed for aquaculture animals contains a very high content of protein; in general, the crude protein content in finfish feedstuff is 25–30% [8] and higher for crustacean animals, which is even up to 40–45% for shrimp species like white-leg shrimp [4]. However, the utilization efficiency of those feeds in water is very low. When feed is added to water, only 25% of protein nitrogen in feedstuff is assimilated to body growth of aquatic animals, and the rest of about an approximate 75% proportion will lose into the water body, via directly excreting as metabolic ammonia from gill, evacuating as urea and feces by cloaca system, or dissolving as other organic nitrogen compounds [9], which are further degraded as inorganic ammonia by microorganisms with hydrolysis enzymes.

4. The main routes for ammonia transformation in aquaculture

There are three routes for ammonia removal or transformation in aquaculture system: intake by photoautotrophic algae, nitrification and nitration of autotrophic nitrobacteria, and assimilation of heterotrophic bacteria [10].

4.1 Route 1: photoautotrophic intake by algae or phytoplankton

Actually, the intake route of ammonia by photoautotrophic algae is the process of well-known photosynthesis as follows [10]:

$$16NH_4^+ + 92CO_2 + 92H_2O + 14HCO_3^- + HPO_4^{2-} \rightarrow C_{106}H_{263}O_{110}N_{16}P + 106O_2 \quad (3)$$

Or when nitrate is as the nitrogen source

$$16NO_3^- + 124CO_2 + 140H_2O + HPO_4^{2-} \rightarrow C_{106}H_{263}O_{110}N_{16}P + 138O_2 + 18HCO_3^- \quad (4)$$

where $C_{106}H_{263}O_{110}N_{16}P$ represents the stoichiometric formula for algae.

In this process, the ionic ammonia of NH_4^+ is the first-order utilized inorganic nitrogen for synthesis of organic materials. However, a carbon to nitrogen to phosphorus ratio (C:N:P) of about 106:16:1 is also needed, indicating that to promote ammonia assimilation, exogenous additions of inorganic carbon and phosphorus sources are needed and that in general make the growth of algae, especially blue-green algae or cyanobacteria, to be very difficult to control and easily result in cyanobacteria blooming, a serious deterioration of water quality and a disaster for human daily life.

4.2 Route 2: autotrophic oxidation by nitrobacteria

Autotrophic nitrobacteria, the chemical autotrophic bacteria, can oxidize ammonia by using inorganic carbon sources without the need of phosphorus [10]:

$$\begin{aligned} NH_4^+ + 1.83O_2 + 1.97HCO_3^- &\rightarrow 0.0244C_5H_7O_2N \\ &+0.976NO_3^- + 2.9H_2O + 1.86CO_2 \end{aligned} \quad (5)$$

where $C_5H_7O_2N$ represents the chemical formula for microbial biomass.

However, the growth rate of nitrobacteria is very low when compared to heterotrophic bacteria, which in turn leads to a low oxidized rate for ammonia. There are also no other efficient supplemental approaches to accelerate this process, which mainly relies on the natural development of nitrobacteria. Furthermore, an intermediate product of this process, nitrite or NO_2^-, another toxic inorganic nitrogen compound for aquaculture animals, would be produced. Nitrite is an unstable product with high oxidized ability comparable to oxygen and thus will oxidize Fe^{2+} in the center of hemoglobin to Fe^{3+}. As a result, oxygen could not combine to hemoglobin and transport to tissues, and thus animals will be asphyxiated, even though there is enough oxygen dissolved in water body [11]. Moreover, the oxidization of ammonia by nitrobacteria would cause numerous accumulation of nitrate (NO_3^-), another inorganic nitrogen compound which could be easily taken by phytoplankton, indicating a potential risk of algae blooming [10, 11]. Finally, the nitrification process could affect water quality, such as exhausting carbonate alkalinity (HCO_3^-) and resulting in reduction of water pH [10].

4.3 Route 3: assimilation by heterotrophic bacteria

Ammonia also could be assimilated by heterotrophic bacteria through a process different from those of photoautotrophic algae (route 1) and autotrophic nitrobacteria (route 2) [10]:

$$NH_4^+ + 1.18C_6H_{12}O_6 + HCO_3^- + 2.06O_2 \rightarrow C_5H_7O_2N + 6.06H_2O + 3.07CO_2 \quad (6)$$

where $C_5H_7O_2N$ represents the chemical formula for microbial biomass like route 2, or Eq. (5). Compared to route 2, sufficient dissolved oxygen is needed for the processing of bio-reaction of Eq. (6) as well, but about half of HCO_3^- will be exhausted. Differently, in Eq. (6) of route 3, carbohydrate ($C_6H_{12}O_6$) is needed, and about 40 times microbial biomass is produced.

5. Novel solution for ammonia assimilation and reuse in aquaculture based on route 3: biofloc technology (BFT)

Ammonia accumulation is the head issue faced in aquaculture, and there are several routes referring to ammonia clearance mentioned above. However, routes 1 and 2 are all not suitable to apply in aquaculture. For route 1, intake of phytoplankton or algae might produce a large number of algae exceeding the biological capacity of water body, and those planktons will be old and die quickly and release toxins harmful to aquatic animals. In regard to route 2, it is mainly applied for effluent treatment in sewage plant, which needs inferior procedures of wastewater, and thus is not suitable in aquaculture as well. Fortunately, according to the principles of route 3 displayed in Eq. (6), a novel technology, in generic nicknamed as biofloc technology (BFT), is developed for aquaculture in recent years, to be used as effectively and environmental-friendly for transforming of ammonia.

5.1 Principal operations of BFT

In accordance with Eq. (6), existing of carbohydrate will promote assimilation of ammonia, companied with synthesis of microbial biomass. However, the content of carbohydrate or C:N in aquaculture water body is lower than the need for bio-reaction of Eq. (6) in general. Although the C:N of bacterial cell composition is about 5:1 [12], it needs a C:N of 15:1 for blooming growth of heterotrophic bacteria to assimilate ammonia [13, 14]. In aquaculture water body, the carbohydrate is mainly from feedstuff added in [7], whose content is usually inadequate for blooming growth of heterotrophic bacteria. For example, taking white-leg shrimp feed usually used in China into consideration, the contents of ingredients, such as crude protein, lipid, fiber, ash, and moisture, are 40, 5.0, 5.0, 15, and 12%, respectively, indicating a calculated C:N of approximate 6:1 according to the relationship between contents of carbohydrate and feed ingredients [15, 16]:

$$\text{Carbon\%} = 0.80 \times \text{Lipid\%} + 0.53 \times \text{Protein\%} + 0.42 \times \text{Carbohydrate\%} + 0.42 \times \text{Fiber\%} \quad (7)$$

$$\text{Carbohydrate\%} = 100 - (\text{Protein} + \text{Lipid} + \text{Fiber} + \text{ash} + \text{moisture})\% \quad (8)$$

Therefore, additional exogenous organic carbon source containing carbohydrate ($C_6H_{12}O_6$) should be supplemented to prompt assimilation of ammonia by improving growth of heterotrophic bacteria, and this is one of the two principal operations for BFT [17].

The other principal operations for BFT are aeration and treatment of by-product. Known from Eq. (6), a huge number of dissolved oxygen is needed to assimilate ammonia by heterotrophic bacteria, and also massive bacteria biomass is produced as by-product, which needs to be treated.

5.2 Addition of carbohydrate

For assimilating 1 mole of ammonia, 1.18 mole of carbohydrate is exhausted according to Eq. (6), which indicates that when 1 g of NH_4^+ exists in water, about 12 g of $C_6H_{12}O_6$ should be added [10, 17]. This needs to supervise the ammonia concentration of water continuously, which is difficult to implement actually. Thus, a general manipulation is that carbon source is added only when ammonia concentration excesses 1 mg/L with the NH_4^+ to $C_6H_{12}O_6$ ratio (w:w) of 1:12 [10, 17]. Of course, the content of carbohydrate contained in material used as carbon source should be determined.

Another way for addition of carbohydrate to improve the bio-reaction of route 3 is adjustment of the C:N in water in real time. For this purpose, the contents of nitrogen in water are determined actually, and then materials rich in carbon or carbohydrate are added to adjust C:N. However, in fact, many times, the adjustment of C:N is not based on the actual carbon and nitrogen concentrations. Alternatively, only when feedstuff is fed, carbon source is considered to add, and the weight for addition is calculated according to the nitrogen content in feedstuff with a C:N of 15:1 [13].

Many materials could be used as carbon source for BFT system, such as acetate [18], glycerol [18], dextrose [19–21], cassava meal [22], cellulose [23], corn flour [24, 25], glucose [18], molasses [26–28], tapioca [29], wheat flour [28, 30], rice flour [16, 30], wheat bran [25, 31], rice bran [20, 29], starch [28, 32], poly-β-hydroxybutyrate (PHB) [33, 34], brewery residues [22], and sugar [32].

5.3 Aeration

The process of ammonia assimilation via heterotrophic microorganisms needs a huge number of oxygen, because of (1) oxygen consumption by respiration of blooming growth of bacteria and (2) oxidized fermentation of organic materials secreted by bacteria [17]. Thus, it is needed to usually equip a robust air blower to blow air into the water body to maintain a highly dissolved oxygen level in water [2], in general at least 5 mg/L [4]. In some cases, even pure oxygen is used for this purpose.

5.4 Treatment of by-product

A result induced by blooming growth of heterotrophic bacteria is substantial accumulation of suspended solids or bioflocs, one of the side effects of utilizing BFT. In a BFT system constructed by the author in the present article, bacteria secrete massive metabolic materials such as protein and polysaccharide, which could bond feeds, feces, debris, and other organic matters together, to become bioflocs and suspended in water under aeration condition (**Figures 2** and **3**). The author also found that sometimes the total suspended solid (TSS) content in BFT system would accumulate to above 800 mg/L (**Figure 3b**). That high level of TSS will be harmful to aquatic animals, which would lead to oxygen depletion, obstruction of fish or shrimp gills, and mortality due to asphyxiation [35]. Therefore, treatment of those accumulated TSS is an important operation for BFT [36].

There are three ways used for treating of TSS. The first one is in situ eaten by fish or shrimp as supplemental food [37, 38] which is also the most frequently used method. The second one is equipping a settling chamber to remove excessive solids [39]. And the last one is using separation systems for biofloc production and aquatic

Figure 2.
Sedimentation of biofloc in an Imhoff cone.

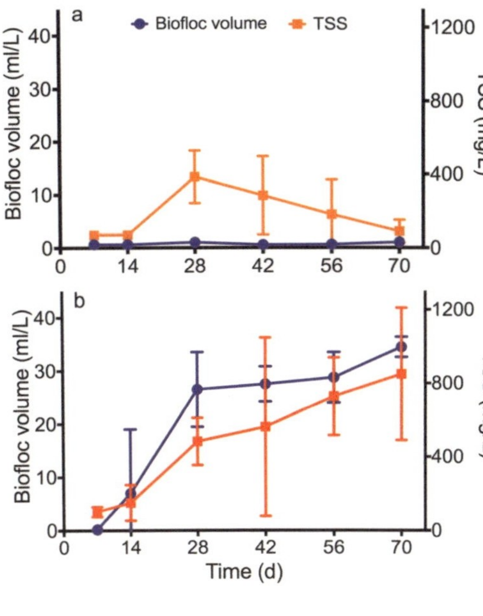

Figure 3.
Productivity characteristics of biofloc volume and TSS in closed traditional system (a) and BFT system (b) over time. Biofloc volume is defined as the volume of sinkable matter in 1 l water placed into an Imhoff cone in 15 min. TSS is represented as the mass (mg) of dry matter in 1 l water after filtering with a 0.45 μl membrane.

animal production, respectively, so that the increasing TSS produced in biofloc production system will not affect the growth of fish or shrimp raised in another system whose water quality should remain controlled by the former system. And for this purpose, four 10 m^3 composite tanks in general need to be fixed for water treatment of 12 tanks with a volume of 500 L per tank [40].

6. Management of dissolved oxygen and ammonia in BFT system

6.1 Dissolved oxygen level

Although numerous amounts of oxygen will be consumed by respiration of a large number of flourishing heterotrophic bacteria, the author in this article supervised that the dissolved oxygen continuously sustained a high level in fact in a *L. vannamei* BFT system equipping air blower, such as root blower (**Figure 4**). In general, air blew into per minute with 1–5% of water volume is adequate to maintain a high dissolved oxygen level. For example, in a pond reserving 1000 m^3 water, the flow rate of root blower should be 10–50 m^3/min.

6.2 Ammonia assimilation efficiency

The speed of ammonia assimilation in BFT system is very fast; Avnimelech [13] reported that ammonia added to water body with a final concentration of 10 mg/L disappeared over a period of about 2 h post addition of glucose as carbon source. The author of the present chapter found that in the BFT system culture *L. vannamei*, a low ammonia level averaging 0.78 mg/L lasted during the whole period of culture in water body (**Figure 5b**). Moreover, nitrite and nitrate concentrations would also be limited at low levels in this BFT system (**Figure 5b**). Conversely, in the control system without water exchange, the average ammonia concentration was up to 6.35 mg/L, which in turn resulted in a very high level of nitrite of 11.78 mg/L subsequently (**Figure 5a**). Consequently, a very high shrimp mortality rate of 70% was found in this control water body, very far higher than that of the BFT system without mortality over the whole experimental period of 70 d (data not shown). Cardona et al. [41] reported that the survival rate of *Litopenaeus stylirostris* was 27.2% higher in the BFT system than that of the conventional system with huge water exchange, which was also contributed to the good water quality represented by those low ammonia and nitrite levels.

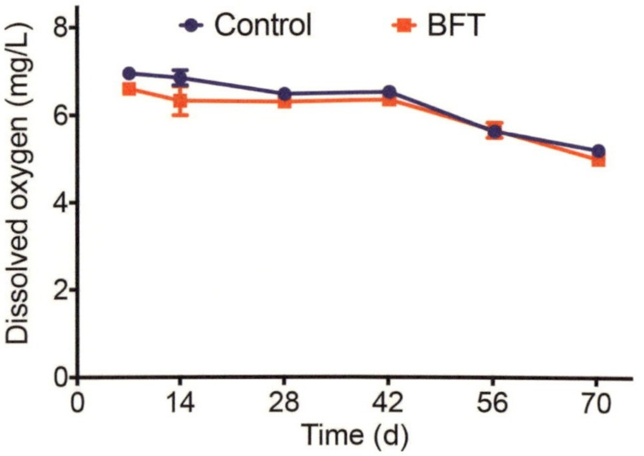

Figure 4.
Dissolved oxygen level in a BFT system over time.

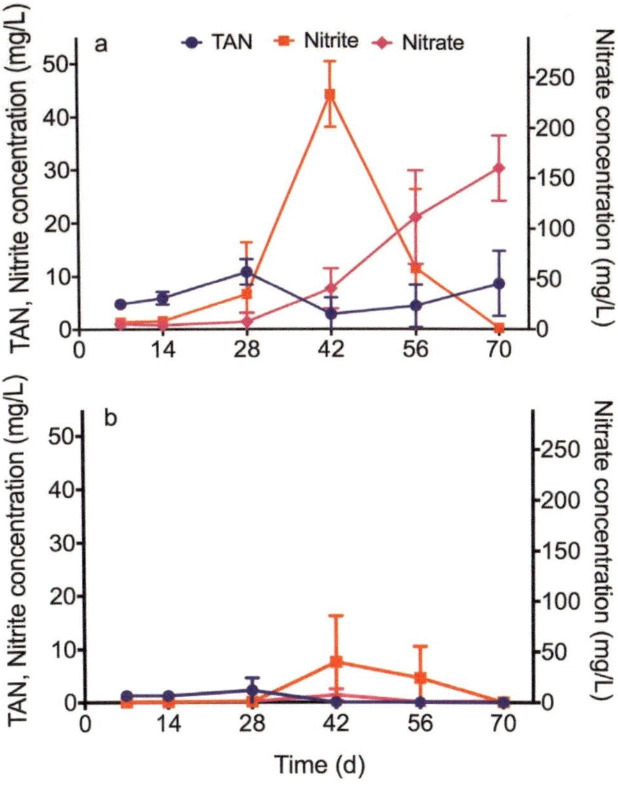

Figure 5.
TAN, nitrite, and nitrate concentrations in closed traditional system (a) and BFT system (b) at different time points.

7. Approaches for reusing of biofloc as supplemental food

Treatment of suspended solids or bioflocs is one of the most important operations for using BFT. Usually, those solids or bioflocs are not removed just as a waste from water. In contrast, they are reused as a complemented food source for aquatic animals, especially omnivorous species such as shrimp and tilapia, in a system adopting BFT. During development of bioflocs, bacteria secrete protein and polysaccharide, which bond with feeds, feces, debris, and other organic matters together. Furthermore, the author of this article found that biofloc was also a nutritional resource that could attract zooplanktons to prey, such as protozoa, rotifers, nematodes, ciliates, and flagellates (**Figure 6**), which in turn provides live and fresh food rich in protein for fish and shrimp.

7.1 Three ways for food use of biofloc

There are three ways for biofloc used as food in aquaculture currently: (i) as a complemented food for fish or shrimp in situ [42–45], (ii) as a gradient for feedstuff to replace fishmeal [46–49], and (iii) as a normal feed to replace partial artificial feedstuff [50–55]. In brief, fish and shrimp consume biofloc rich in microbe, phytoplankton, and zooplankton as a food directly. Because animals take biofloc as a vice food source, thus in fact the biofloc hunted by fish and shrimp is only a few parts of the whole. In other words, most of the bioflocs remain in the water body, which may be an obstacle for growth of fish and shrimp. For alleviating the negative effect of biofloc on animal growth, excessive parts should be collected from water body and could be taken as an alternative protein source for preparing feedstuff. Even more, biofloc is fed to fish or shrimp as feedstuff directly due to its whole and high nutritional value.

7.2 Factors influencing nutritional value of biofloc

Nutritional value of biofloc is important for its reuse. However, this value is affected by several factors. Because the development of biofloc is sponsored and prompted by accumulation of ammonia and addition of carbon source, it is suspected that feedstuff and carbon source [30], especially the last one, would impressively affect biofloc nutritional composition and value (**Table 1**). For example, protein content and oil content of feedstuff will affect those of in biofloc. With regard to carbon source, there are two main types of carbon sources: (i) simple structure carbon sources with easily dissolving ability in water, such as glucose, sucrose, and sodium acetate [17, 20], and (ii) complex compounds, like flour or bran of rice and wheat [56] and brewery residues, which are a by-product from beer production industry [22]. In general, complex carbon source is more difficult to dissolve and more powerful in improving biofloc nutritional value, which in turn improves the growth of fish or shrimp [17, 20]. This carbon source is not easy to degrade with big diameter so that animals in water could easily prey and eat them directly; thus, except for being used as carbohydrate, those materials also contain other nutritional materials essential for growth of fish and shrimp, such as proteins, oils, vitamins, and minerals, even carotenoids [57, 58].

7.3 Applications of biofloc as complemented food in aquaculture

BFT has been successfully used for culture of fish and shrimp, such as tilapia, carp, and *L. vannamei*. Results showed that BFT increased individual fish weight,

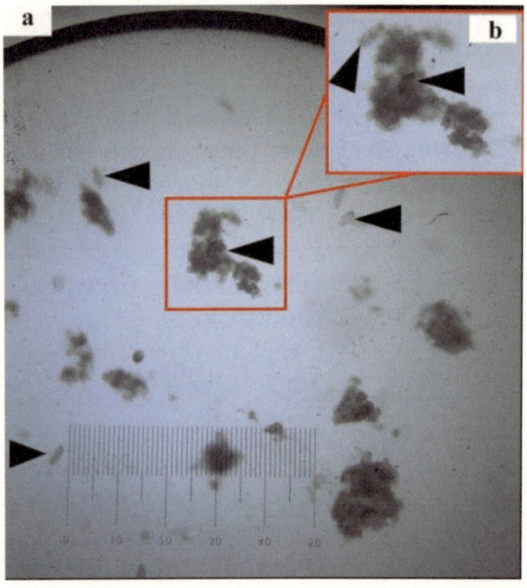

Figure 6.
Bioflocs observed with a light microscope. The minimal scale of the rule in the down part of the figure represents 25 μm. Arrows indicate free zooplanktons (a), and zooplanktons prey food from biofloc (b).

Animals	Carbon sources	Feedstuff composition		C:N	Biofloc composition		Ref.
		CP	CL		CP	CL	
L. vannamei	Molasses and wheat bran	42.5	–	20:1	28.7–43.1	2.11–3.62	[59]
	Sucrose	35–40	7–9	–	24.01	3.31	[60]
	Molasses	38	9	15:1	27.43	0.86	[28]
	Starch				23.1	1.14	
	Wheat flour				30.73	2.18	
	Mixture of molasses, starch, and wheat flour with equal weight ratio				25.46	1.24	
Tilapia	Wheat flour	24	6.23	–	37.93	3.16	[61]
		35	6.24		38.41	3.23	
	Feed	22	12.3	11.6:1	50.6	2.6	[62]
		35	119	8.4:1	53.5	1.9	
	Poly-β-hydroxybutyric	30	4	–	34.06	6.58	[34]
	Glucose				38.53	6.06	
Farfantepenaeus paulensis	Molasses and wheat bran	40	13.1	20:1	30.4	0.47	[37]
		35	–	–	18.4	0.3	[63]
Labeo rohita	Wheat flour	29.6–35.4	4.2–16.5	10:1	35.4	1.1	[52]
Penaeus monodon	Wheat flour	40	–	10:1	24.3	3.53	[53]
Catfish	Glycerol	43	6	10:1	44.27	5.84	[64]
				15:1	38.65	7.35	
				20:1	32.64	10.78	
Green cucumber	Glucose	20.37	2.45	15:1	32.29	4.19	[65]
	Sucrose				28.04	4.30	
	Starch				21.67	3.83	
White cucumber	Glucose				27.27	4.25	
	Sucrose				27.48	3.89	
	Starch				21.23	3.76	

Note: CP, crude protein; CL, crude lipid.

Table 1.
Compositions of biofloc produced from different feedstuff and carbon sources.

weight gain, and protein efficiency ratio of tilapia by 12.54, 9.46, and 22.2%; decreased feed conversion rate (FCR) by 17.5% [66]; elevated total weight gain and specific growth rate increase by 128% and 112% [67], respectively; and also significantly increased the final weight, weight gain, and length of *L. vannamei* [37, 63]. And those results for enhancing growth of fish and shrimp are mainly contributed to biofloc. Research from Tierney and Ray [2] revealed that the shrimp in the BFT treatment received an estimated 87% of their carbon and 66% of their nitrogen from the pelleted feed source, whereas the rest of 13% and 34% came from the biofloc, respectively.

When bioflocs were eaten directly by fish or shrimp, the protein utilization efficiency of feed elevated by 29% [68], and the FCR decreased by about 18% for tilapia [66, 67], and also decreased for *Penaeus monodon* [69]. It is found that the protease, lipase, and amylase activities in the intestine, liver, and stomach of tilapia [66, 67] and *L. vannamei* were all significantly raised [25], which improves the digestion and intake of feed. A large number of heterotrophic bacteria are located in biofloc, most of which will secret enzymes like protease, lipase, and amylase, indicating that preying on biofloc would directly increase the content of those enzymes in digestive organs of fish and shrimp. Those bacteria also would secret digestive enzymes in vivo after entering, colonizing and propagating in the intestine [25]. In a BFT system for culturing *L. vannamei*, the feedstuff containing 35% of protein could be altered with feedstuff whose protein content decreased to 25%, because most nutrition materials lacking in low protein content feed could be compensated from biofloc. In another way, about 25% feedstuff could be saved when culturing *L. vannamei* with BFT because biofloc was hunted as a food in situ [54]. It is even reported that no feed should be added in some BFT systems for culturing *L. vannamei* and *Macrobrachium rosenbergii* [18, 37].

Biofloc is also used as an alternative protein source for fishmeal sometimes. The protein content in biofloc is evidenced to be very high, in general 25–40% [34, 63, 66, 70], in a case even up to 50% [62]. The essential amino acids were also rich in biofloc, and its composition was also highly in agreement with that in the fish body [71], indicating that it is valuable for growth of fish and shrimp. Dantas et al. [46] and Kuhn et al. [48] replaced 30% of fishmeal or soy meal with biofloc to manufacture feedstuff for feeding of *L. vannamei*, and results showed that the growth of shrimp is not affected. Kuhn et al. [47] thought that the whole soy meal could be replaced with biofloc and 67% for fishmeal. However, only appropriate replacement of fishmeal with biofloc could prompt the growth of shrimp obviously, approximately 15.16–16.5% [49].

In some cases, biofloc was collected and dried to make pellets and then fed to fish or shrimp like artificial or formulated feedstuff. Carps, *Labeo rohita*, and *L. vannamei* was cultured successfully with biofloc pellets according to a replacement ratio of 25, 50, and 50% to formulated feedstuff, respectively, without any negative effects on their growth [50–52]. However, when feedstuff was replaced 35% with biofloc, the growth of *L. vannamei* appears to be the fastest [72]. In fact, a replacement ratio lower than 10% could improve the growth of animals [55]. For example, when the replacement ratio was 4 and 8%, the individual mean body weight of *P. monodon* increased by 16.9 and 13.9%, respectively [53].

8. Prospects

Meeting future demand for fish is very important for global food security. However, barriers to growth have to be explicitly recognized to the environmental and economic pillars of sustainability [73]. Fortunately, BFT could fulfill those requests for sustainable development of aquaculture.

8.1 Environmental advantages of BFT for sustainable development of aquaculture under framework of the FAO

Except availability of land and water, environmental impact is another possible main constraint to aquaculture growth. Thus, aquaculture systems that reduce eutrophication risks and other environmental costs while providing income and extended social benefits should be developed [73]. For this purpose, the FAO thinks that herbivorous and omnivorous species should be promoted and integrated aquaculture including multitrophic aquaculture is also an alternative, in which by-products (wastes) from one species are recycled to become inputs (fertilizers, food, and energy) for another [73].

From this point of view, in practical aquaculture operations, BFT utilizes by-products from agriculture industries, such as cassava meal [22], molasses [26–28], tapioca [29], wheat bran [25, 31], rice bran [20, 29], and brewery residues [22], as fertilizers for assimilating organic and inorganic materials. And in turn, its own by-product, biofloc, becomes complemented food for aquatic suspension or deposit feeders, like herbivorous fish. Some omnivorous aquatic animals, such as shrimp and tilapia, were all very suitable to be cultured with BFT [44, 62]. Due to the characteristics of in situ treatment of water quality and supplying of organic biofloc food, aquaculture systems that adopted BFT only need a few water exchange, even no water exchange, and decrease artificial feedstuff inputs, indicating reduced eutrophication risks of environment and the use of wild fish for aquaculture feeds to reserve balance of ecosystem.

8.2 Practical application of BFT for economical benefits

Rego et al. [74] analyzed the financial viability of inserting the BFT system (625 m^2 each pond) and maintaining the conventional culture system (2.86 ha each pond) for the marine shrimp *L. vannamei* in a farm located in the state of Pernambuco, northeastern Brazil. The total production costs of BFT were eight times higher than the conventional system. However, operating profit and profitability index were US$ 51,871.54 per hectare per year and 30.22% for BFT, and US$ 21,523.83 and 59.79% for the conventional system, respectively. In an investment analysis, indicators were favorable for both systems, with greater expressiveness of the net present value (NPV) for the BFT (US$ 142,004.42) and internal rate of return (IRR) 4.5 times higher for conventional system (131.86%). When the cash flows were designed for 10 years to the discount rates of 10, 13, and 16%, the BFT system showed greater sensitivity to changes in rates, reducing significantly the NPV when interest rates increased. Risk was only observed in the BFT system, with up to 15% of probability when subjected to the discount rate of 16%. Both shrimp production systems represent a significant investment alternative for the rural sector in northeastern Brazil, because even from the perspective of risk management, the IRR has 90% probability ranging from 7.66 to 59.40% for the BFT and from 67.96 to 201.03% for the conventional systems [75].

8.3 Further improvements for reuse of biofloc

Undoubtedly, BFT is a novel solution for transformation of ammonia in aquaculture. However, how to effectively reuse or deposit biofloc, the by-product of assimilation of ammonia in BFT system in situ, as a supplemental food for aquatic animals, needs more researches in detail.

The consumed efficiency of biofloc by fish or shrimp in situ is not adequate high, resulting in gradual accumulation of TSS in BFT system because of huge numerous organic materials produced by blooming growth of heterotrophic bacteria. Thus, the causes contributed to this low efficiency should be researched. Furthermore, the strategies for improving the utilization efficiency of biofloc should be assessed as well, such as improving accumulation of lipid of biofloc, which will increase the nutritional value. Usually, the total lipid content in biofloc is too low to be sufficient for demand of fish and shrimp (**Table 1**). Previous studies found that the lipid contents of biofloc were 0.5–0.6% [63, 70], 1.03% [66], or 4.0% [62], respectively, which were all lower than the demands for lipid of aquatic animals [8]. For example, the recommended total lipid level in the diet for shrimp is in general higher than 6.5% [4]. Although external equipment could be used to settle the excessive part of biofloc, how to treat this deposit containing high content organic matter and bacteria, part of which may be pathogens, was also a problem [39].

The efficiency for producing biofloc also needs to be elevated, if biofloc is used as a gradient of formulated feedstuff for replacement of fishmeal or soybean meal or used to feed to aquatic animals directly as a food with whole nutritional gradients usually contained in artificial feedstuff. The productivity of biofloc recent is not adequate for those uses in practical operations.

Moreover, the improvement of biofloc palatability should be researched, which is important to the utilization of biofloc either eaten in situ or used as a food source [76]. Attractants or feeding promoting agents, like garlicin, betaine (trimethylglycine), trimethylamine oxide (TMAO), and s,s-dimethyl-β-propionic acid thetine (DMPT), could be taken into consideration as additives during development of biofloc in situ or preparing process for biofloc pellets. Thus, the effects of those agents on biofloc attraction to fish and shrimp should be studied in detail, respectively.

9. Conclusions

Biofloc technology (BFT) supplies a novel solution for this issue without huge water exchange, even zero water exchange. In general, ammonia would be removed quickly within several hours in a BFT system. Moreover, because of the very high nutritional value for fish and shrimp, bioflocs, the by-product of BFT, could also be reused as a complemented food in situ or a gradient for feedstuff to replace expensive fishmeal, and biofloc also could be processed to formulate diet to feed fish and shrimp directly. However, some aspects with regard to the effective use of biofloc as a food source for fish and shrimp, such as high lipid content, productivity, and palatability, need to be further researched in detail.

Acknowledgements

This work is supported by the development funds of the Chinese central government to guide local science and technology (2017CT5013); the Sci-Tech program of Hunan province, China (2016NK2132); and the science and research program of the Education Department of Hunan province, China (16C1085, 18B394).

Conflict of interest

The author declares no conflict of interest.

Notes/thanks/other declarations

The author also thanks the support from Collaborative Innovation Center (Hunan) for Efficient and Health Production of Fisheries, Hunan Engineering Research Center of Aquatic Organism Resources and Environmental Ecology, Hunan Engineering Research Center of Aquatic Organism Resources and Environmental Ecology, and Academician Workstation (Fisheries) of Hunan Province.

Appendices and nomenclature

BFT	biofloc technology
C:N (P)	carbon to nitrogen (to phosphorus) ratio
CL	crude lipid
CP	crude protein
DMPT	s,s-dimethyl-β-propionic acid thetine
FAO	Food and Agriculture Organization
FCR	feed conversion rate
IRR	internal rate of return
LC_{50}	median lethal concentration
NPV	net present value
NRC	National Research Council
PHB	poly-β-hydroxybutyric
ppm	part(s) per million
TAN	total ammonia nitrogen
TMAO	trimethylamine oxide
TSS	total suspended solids

Author details

Hai-Hong Huang
College of Life and Environmental Science, Hunan University of Arts and Science, Key Laboratory of Health Aquaculture and Product Processing in Dongting Lake Area, Zoology Key Laboratory of Hunan Higher Education, Changde, China

*Address all correspondence to: shinkanh@nwsuaf.edu.cn

References

[1] FAO. The State of World Fisheries and Aquaculture. Rome; 2018

[2] Tierney TW, Ray AJ. Comparing biofloc, clear-water, and hybrid nursery systems (Part I): Shrimp (*Litopenaeus vannamei*) production, water quality, and stable isotope dynamics. Aquacultural Engineering. 2018;**82**: 73-79

[3] Lin Y, Chen J. Acute toxicity of ammonia on *Litopenaeus vannamei* Boone juveniles at different salinity levels. Journal of Experimental Marine Biology and Ecology. 2001;**259**:109-119

[4] Van Wyk P et al. Farming Marine Shrimp in Recirculating Freshwater Systems. Florida Department of Agriculture and Consumer Services: Tallahassee; 1999

[5] Lekang OI. Aquaculture Engineering. 2nd ed. Willey-Blackwell: Chichester, UK; 2013

[6] Mosquera-Corral A, Campos FGL, Mendez R. Partial nitrification in a SHARON reactor in the presence of salts and organic carbon compounds. Process Biochemistry. 2005;**40**:3109-3118

[7] Fungesmith SJ, Mrp B. Nutrient budgets in intensive shrimp ponds: Implications for sustainability. Aquaculture. 1998;**164**:117-133

[8] NRC. Nutrient requirements of fish and shrimp. In: Animal Nutrition Series. Washington, DC: The National Academies Press, National Research Council of the National Academies; 2011

[9] Piedrahita RH. Reducing the potential environmental impact of tank aquaculture effluents through intensification and recirculation. Aquaculture. 2003;**226**:35-44

[10] Ebeling JM, Timmons MB, Bisogni JJ. Engineering analysis of the stoichiometry of photoautotrophic, autotrophic, and heterotrophic removal of ammonia–nitrogen in aquaculture systems. Aquaculture. 2006;**257**:346-358

[11] Noga EJ. Fish Disease Diagnosis and Treatment. 2nd ed. Ames: Wiley-Blackwell; 2010

[12] Rittmann BE, McCarty PL. Environmental Biotechnology: Principles and Applications. New York: McGraw-Hill; 2002

[13] Avnimelech Y. Carbon/nitrogen ratio as a control element in aquaculture systems. Aquaculture. 1999;**176**:227-235

[14] Asaduzzaman M et al. C/N ratio control and substrate addition for periphyton development jointly enhance freshwater prawn *Macrobrachium rosenbergii* production in ponds. Aquaculture. 2008;**280**:117-123

[15] Hari B et al. Effects of carbohydrate addition on production in extensive shrimp culture systems. Aquaculture. 2004;**241**:179-194

[16] Kumar S et al. Effects of biofloc under different carbon sources and protein levels on water quality, growth performance and immune responses in black tiger shrimp *Penaeus monodon* (Fabricius, 1978). Aquaculture Research. 2017;**48**:1168-1182

[17] Avnimelech Y. Biofloc Technology: A Practical Hand Book. 3rd ed. The World Aquaculture Society: Baton Rouge, Louisiana, EUA; 2015

[18] Crab R et al. The effect of different carbon sources on the nutritional value of bioflocs, a feed for *Macrobrachium rosenbergii* postlarvae. Aquaculture Research. 2010;**41**:559-567

[19] de Lorenzo MA et al. Intensive hatchery performance of the Pacific white shrimp in biofloc system. Aquacultural Engineering. 2015;**67**: 53-58

[20] Serra FP et al. Use of different carbon sources for the biofloc system adopted during the nursery and grow-out culture of *Litopenaeus vannamei*. Aquaculture International. 2015;**23**: 1325-1339

[21] Suita SM et al. Dextrose as carbon source in the culture of *Litopenaeus vannamei* (Boone, 1931) in a zero exchange system. Latin American Journal of Aquatic Research. 2015;**43**: 526-533

[22] Sena Fugimura MM et al. Brewery residues as a source of organic carbon in *Litopenaeus schmitti* white shrimp farms with BFT systems. Aquaculture International. 2015;**23**:509-522

[23] Deng M et al. The effect of different carbon sources on water quality, microbial community and structure of biofloc systems. Aquaculture. 2018;**482**: 103-110

[24] Asaduzzaman M et al. Effects of C/N ratio and substrate addition on natural food communities in freshwater prawn monoculture ponds. Aquaculture. 2010;**306**:127-136

[25] Wang C et al. Effects of different carbon sources addition on nutrition composition and extracellular enzymes activity of bioflocs, and digestive enzymes activity and growth performance of *Litopenaeus vannamei* in zero-exchange culture tanks. Aquaculture Research. 2016;**47**: 3307-3318

[26] de Souza DM et al. Use of molasses as a carbon source during the nursery rearing of *Farfantepenaeus brasiliensis* (Latreille, 1817) in a biofloc technology system. Aquaculture Research. 2014;**45**: 270-277

[27] do Espirito Santo CM et al. Soybean molasses as an organic carbon source in the farming of *Litopenaeus vannamei* (Boone, 1931) in a biofloc system. Aquaculture Research. 2017;**48**: 1827-1835

[28] Khanjani MH et al. Nursery performance of Pacific white shrimp (*Litopenaeus vannamei* Boone, 1931) cultivated in a biofloc system: The effect of adding different carbon sources. Aquaculture Research. 2017;**48**: 1491-1501

[29] Ekasari J et al. Immune response and disease resistance of shrimp fed biofloc grown on different carbon sources. Fish & Shellfish Immunology. 2014;**41**: 332-339

[30] Panigrahi A et al. Carbohydrate sources deferentially influence growth performances, microbial dynamics and immunomodulation in Pacific white shrimp (*Litopenaeus vannamei*) under biofloc system. Fish & Shellfish Immunology. 2019;**86**:1207-1216

[31] Zhao D et al. Effects of different carbon sources on bioactive compound production of biofloc, immune response, antioxidant level, and growth performance of *Litopenaeus vannamei* in zero-water exchange culture tanks. Journal of the World Aquaculture Society. 2016;**47**:566-576

[32] Bakhshi F et al. Use of different carbon sources for the biofloc system during the grow-out culture of common carp (*Cyprinus carpio* L.) fingerlings. Aquaculture. 2018;**484**:259-267

[33] da Silva BC et al. Dietary supplementation with butyrate and polyhydroxybutyrate on the performance of Pacific white shrimp in biofloc systems. Journal of the World Aquaculture Society. 2016;**47**:508-518

[34] Zhang N et al. Growth, digestive enzyme activity and welfare of tilapia (*Oreochromis niloticus*) reared in a biofloc-based system with poly-beta-hydroxybutyric as a carbon source. Aquaculture. 2016;**464**:710-717

[35] Gaona CAP et al. Effect of different total suspended solids levels on a *Litopenaeus vannamei* (Boone, 1931) BFT culture system during biofloc formation. Aquaculture Research. 2017; **48**:1070-1079

[36] Arantes R et al. A comparison between water exchange and settling tank as a method for suspended solids management in intensive biofloc technology systems: effects on shrimp (*Litopenaeus vannamei*) performance, water quality and water use. Aquaculture Research. 2017;**48**: 1478-1490

[37] Emerenciano M et al. Biofloc technology application as a food source in a limited water exchange nursery system for pink shrimp *Farfantepenaeus brasiliensis* (Latreille, 1817). Aquaculture Research. 2012;**43**:447-457

[38] Hargreaves J. Biofloc Production Systems for Aquaculture. Southern Regional Aquaculture Center; 2013

[39] Ray AJ et al. Suspended solids removal to improve shrimp (*Litopenaeus vannamei*) production and an evaluation of a plant-based feed in minimal-exchange, super intensive culture systems. Aquaculture. 2010;**299**:89-98

[40] Nhi Nguyen Huu Y et al. Comparative evaluation of Brewer's yeast as a replacement for fishmeal in diets for tilapia (*Oreochromis niloticus*), reared in clear water or biofloc environments. Aquaculture. 2018;**495**: 654-660

[41] Cardona E et al. Bacterial community characterization of water and intestine of the shrimp *Litopenaeus stylirostris* in a biofloc system. BMC Microbiology. 2016;**16**

[42] Ekasari J et al. The size of biofloc determines the nutritional composition and the nitrogen recovery by aquaculture animals. Aquaculture. 2014; **426**:105-111

[43] Burford MA et al. The contribution of flocculated material to shrimp (*Litopenaeus vannamei*) nutrition in a high-intensity, zero-exchange system. Aquaculture. 2004;**232**:525-537

[44] Day SB, Salie K, Stander HB. A growth comparison among three commercial tilapia species in a biofloc system. Aquaculture International. 2016;**24**:1309-1322

[45] Monroy-Dosta MC et al. Microbiology community composition and abundance associated to biofloc in tilapia aquaculture. Revista De Biologia Marina Y Oceanografia. 2013;**48**:511-520

[46] Dantas EM Jr et al. Partial replacement of fishmeal with biofloc meal in the diet of postlarvae of the Pacific white shrimp *Litopenaeus vannamei*. Aquaculture Nutrition. 2016; **22**:335-342

[47] Kuhn DD et al. Evaluation of two types of bioflocs derived from biological treatment of fish effluent as feed ingredients for Pacific white shrimp, *Litopenaeus vannamei*. Aquaculture. 2010;**303**:28-33

[48] Kuhn DD et al. Evaluation of bioflocs derived from confectionary food effluent water as a replacement feed ingredient for fishmeal or soy meal for shrimp. Aquaculture. 2016;**454**: 66-71

[49] Valle BCS et al. Replacement of fishmeal by fish protein hydrolysate and biofloc in the diets of *Litopenaeus vannamei* postlarvae. Aquaculture Nutrition. 2015;**21**:105-112

[50] Najdegerami EH, Bakhshi F, Lakani FB. Effects of biofloc on growth performance, digestive enzyme activities and liver histology of common carp (*Cyprinus carpio* L.) fingerlings in zero-water exchange system. Fish Physiology and Biochemistry. 2016;**42**: 457-465

[51] Khatoon H et al. Biofloc as a potential natural feed for shrimp postlarvae. International Biodeterioration & Biodegradation. 2016;**113**:304-309

[52] Mahanand SS, Moulick S, Rao PS. Optimum formulation of feed for rohu, *Labeo rohita* (Hamilton), with biofloc as a component. Aquaculture International. 2013;**21**:347-360

[53] Anand PSS et al. Effect of dietary supplementation of biofloc on growth performance and digestive enzyme activities in *Penaeus monodon*. Aquaculture. 2014;**418**:108-115

[54] Rostika R. The reduction feed on shrimp vaname (*Litopenaues vannamae*) replaced by the addition biofloc in Ciamis District. Research Journal of Biotechnology. 2014;**9**:56-59

[55] Sui LY et al. Increased carbon and nitrogen supplementation in Artemia culture ponds results in higher cyst yields. Aquaculture International. 2013; **21**:1343-1354

[56] Vilani FG et al. Strategies for water preparation in a biofloc system: Effects of carbon source and fertilization dose on water quality and shrimp performance. Aquacultural Engineering. 2016;**74**:70-75

[57] Zhang K et al. Effect of using sodium bicarbonate to adjust the pH to different levels on water quality, the growth and the immune response of shrimp *Litopenaeus vannamei* reared in zero-water exchange biofloc-based culture tanks. Aquaculture Research. 2017;**48**:1194-1208

[58] Ju ZY et al. Determination of microbial community structures of shrimp floc cultures by biomarkers and analysis of floc amino acid profiles. Aquaculture Research. 2010;**39**:118-133

[59] Maicá PF, Borba MRD, Wasielesky W Jr. Effect of low salinity on microbial floc composition and performance of *Litopenaeus vannamei* (Boone) juveniles reared in a zero-water-exchange super-intensive system. Aquaculture Research. 2012;**43**:361-370

[60] Ray AJ, Lotz JM. Comparing salinities of 10, 20, and 30‰ in intensive, commercial-scale biofloc shrimp (*Litopenaeus vannamei*) production systems. Aquaculture. 2017; **476**:29-36

[61] Azim ME, Little DC. The biofloc technology (BFT) in indoor tanks: Water quality, biofloc composition, and growth and welfare of Nile tilapia (*Oreochromis niloticus*). Aquaculture. 2008;**283**:29-35

[62] Azim ME, Little DC, Bron JE. Microbial protein production in activated suspension tanks manipulating C:N ratio in feed and the implications for fish culture. Bioresource Technology. 2008;**99**: 3590-3599

[63] Emerenciano M et al. Effect of biofloc technology (BFT) on the early postlarval stage of pink shrimp *Farfantepenaeus paulensis*: Growth performance, floc composition and salinity stress tolerance. Aquaculture International. 2011;**19**:891-901

[64] Dauda AB et al. Influence of carbon/ nitrogen ratios on biofloc production and biochemical composition and subsequent effects on the growth, physiological status and disease resistance of African catfish (*Clarias*

gariepinus) cultured in glycerol-based biofloc systems. Aquaculture. 2018: S0044848617317131

[65] Chen J et al. Regulation of growth, intestinal microbiota, non-specific immune response and disease resistance of sea cucumber *Apostichopus japonicus* (Selenka) in biofloc systems. Fish and Shellfish Immunology. 2018;**77**:175-186

[66] Long L et al. Effect of biofloc technology on growth, digestive enzyme activity, hematology, and immune response of genetically improved farmed tilapia (*Oreochromis niloticus*). Aquaculture. 2015;**448**:135-141

[67] Luo G et al. Growth, digestive activity, welfare, and partial cost-effectiveness of genetically improved farmed tilapia (*Oreochromis niloticus*) cultured in a recirculating aquaculture system and an indoor biofloc system. Aquaculture. 2014;**422**:1-7

[68] Avnimelech Y, Ritvo G. Shrimp and fish pond soils: processes and management. Aquaculture. 2003;**220**: 549-567

[69] Kumar S et al. Effects of carbohydrate supplementation on water quality, microbial dynamics and growth performance of giant tiger prawn (*Penaeus monodon*). Aquaculture International. 2014;**22**:901-912

[70] Emerenciano M et al. Evaluation of biofloc technology in pink shrimp Farfantepenaeus duorarum culture: Growth performance, water quality, microorganisms profile and proximate analysis of biofloc. Aquaculture International. 2013;**21**:1381-1394

[71] Wei Y, Liao SA, Wang AL. The effect of different carbon sources on the nutritional composition, microbial community and structure of bioflocs. Aquaculture. 2016;**465**:88-93

[72] Xu WJ et al. Preliminary investigation into the contribution of bioflocs on protein nutrition of *Litopenaeus vannamei* fed with different dietary protein levels in zero-water exchange culture tanks. Aquaculture. 2012;**350**:147-153

[73] FAO. The State of World Fisheries and Aquaculture: Opportunities and Challenges. Rome: FAO; 2014

[74] Soares Rego MA et al. Financial viability of inserting the biofloc technology in a marine shrimp *Litopenaeus vannamei* farm: A case study in the state of Pernambuco, Brazil. Aquaculture International. 2017;**25**: 473-483

[75] Soares Rego MA et al. Risk analysis of the insertion of biofloc technology in a marine shrimp *Litopenaeus vannamei* production in a farm in Pernambuco, Brazil: A case study. Aquaculture. 2017; **469**:67-71

[76] Sanhotra MK. Shrimp feed formulation and feed management. CMFRI Special Publication. 1994

Chapter 2

Hybridization in Carps and Early Detection of Carp Hybrids Using PCR-Based Kit

Jayasankar Pallipuram

Abstract

Hybridization is the mating of genetically differentiated individuals or groups and may involve crosses within a species or between separate species. Hybridization can be natural or human-mediated. Reproductive barriers prevent excessive introgression in the former, whereas more often introgression and genetic pollution happen in the latter. Hybridization is more widespread among members of Cyprinidae than any other groups of freshwater fishes. In many carp hatcheries in India, breeders of catla (*Catla catla*) and rohu (*Labeo rohita*) are kept in the same tank for breeding, resulting in production of hybrid seeds. Fish hybrids can pose a serious threat to the aquatic environment biodiversity. Consequently, genetic monitoring of organisms is entailed to unambiguously identify parental species and their hybrids. Adopting a multiplex PCR using β-actin gene primers, a kit has been developed to distinguish between the hybrids from their parental species. Agarose electrophoresis revealed one band of about 100 bp in size specific for rohu, another at 300 bp specific for catla, and both bands in hybrid. The kit was tested successfully with the samples collected from many hatcheries located in four Indian states. The rohu-catla early hybrid identification PCR kit could serve as a stepping stone for carp seed certification and hatchery accreditation.

Keywords: hybridization, introgression, carps, hatchery, PCR

1. Introduction

Hybridization is defined as the mating of genetically differentiated individuals or groups and may involve crosses within a species (also known as line crossing or strain crossing) or crosses between separate species [1]. Issues related to hybridization are complex, making the job of conservation biologists tougher. Interspecific and intergeneric hybridization do happen naturally, and it is considered to play an important role in evolution process [2]. Indian major carps (IMCs) comprising of *Labeo rohita*, *Catla catla*, and *Cirrhinus mrigala*, owing to their fast growth and taste, enjoy a prime position in the Indian aquaculture scenario. These carps attain a marketable size of 800–1000 g in less than a year and are generally propagated on an extensive and/or intensive scale in a polyculture system [3]. Among the three IMCs, catla and rohu are generally chosen for freshwater aquaculture due to their faster growth. IMCs, though originally inhabitants of the Ganga River network in North India and the rivers of Pakistan, Bangladesh, Nepal, and Burma, are also transplanted into other rivers in central as well as peninsular India and also in aquaculture systems of Nepal and Sri Lanka. In 2005 global carp production reached 28.8 million tons, accounting for 37.5% in quantity and 25.6% in value of total aquaculture production [4]. China with 21.05 and India with 3.9 million tons were the top carp producers in the world. Indian major carps catla (2.76 million tons), rohu (2.91 million tons), and mrigal (0.47 million tons) were among 29 species with production over 100 t in 2005 [4].

It was noted that the growth of IMCs affected the grow-out culture phase and the profitability in carp farming is decreasing [5, 6]. The growth of carps is affected largely due to poor quality and mixed seed of carps produced by breeding carps of different species at the same time in the spawning pools of carp hatcheries that make easy hybrids. To make more profit, hatchery managers practice this in breeding programs in hatcheries when there is paucity of brood fish of desired species that is either males or females. This practice has been rampant in many parts of the Indian subcontinent. Unlike other agricultural crops and domestic land animals, the hybrids did not grow better than their native natural parents. Good quality seed of catla and rohu are in great demand in the Indian subcontinent. Seed is one of the indispensable resources needed for aquaculture. Taking the advantage of seed demand of these two species, many hatchery owners and seed producers supply hybrid seed (catla × rohu or rohu × catla) of these two species in the name of pure rohu or catla during the young stages. These hybrids cannot be easily differentiated from each other morphologically at early stages of development, e.g., hatchling and early fry stages. Among the most pressing issues concerning seed in global aquaculture development include inadequate and unreliable supply of quality seed, genetic quality, and inadequate hatchery technology.

For production of quality seed in aquaculture, many methods were tried in the past, including intergeneric and interspecific hybridization. Hybridization technique was used by aquaculturists in the hope of producing aquatic organisms with specific desirable traits or general improvement in performance. Intergeneric and interspecific hybridization programs have been applied in fish farms with the purpose of producing animals that perform better than the parental species (hybrid vigor) [1].

Nevertheless, many species are jeopardized by hybridization and genetic introgression, and these are particularly prevalent threats to the diversity of freshwater fish [7–10]. If fertile, hybrids can genetically contaminate natural and farmed stocks through genetic homogenization. They may also compete in several ways with the native parental lineages [2, 11, 12].

Currently, several genetic markers have been developed for different species that are used in aquaculture programs [13, 14]. PCR-RFLP and multiplex PCR are considered fast, simple, and inexpensive, but they have rarely been used in the characterization of hybrids or in aquaculture in general. These techniques are important tools in species identification, especially in studies related to the biological conservation and forensic identification [15, 16]. Natural occurrence of both interspecific and intergeneric hybrids of Indian major carps has been reported mostly from reservoirs and other natural ecosystems. From natural ecosystems such as reservoir and dry bundhs, several hybrids have been recorded [17]. Many of these hybrids were found to be intermediate in characters of the parent species. Only a few hybrids, both artificially produced and naturally occurring, have been studied in detail for their cultural qualities and adaptability to various environments. Several interspecific and intergeneric hybrids of Indian major carps are *Catla catla, Labeo rohita, Cirrhinus mrigala*, and *Labeo calbasu* [18, 19], and those of Indian major carps with exotic carps, viz., common carp [20] and silver carp [21], have been artificially produced through hypophysation. These hybrids were not popular due to poor survival and many undesired traits in aquaculture.

Quality seed is a fundamental prerequisite for sustainable and successful aquaculture, be it small-scale or commercial farming. Inadequate supply of quality seed is often suggested as a major constraint for aquaculture in many parts of the world [22]. The issue of quality comes to the attention of producers only after a certain period of time when performance indicators (e.g., growth, production, survival, and disease) consistently point a finger towards seed quality. Factors which contribute to production of poor quality seed would have become established as a normal practice in the system.

Field level morphological identification of seeds of commercially important carp species, viz., catla and rohu, at early stage is a difficult task as experience has shown. Many times fish farmers get cheated by unethical practices by hatchery owners as they sell these hybrids in the name of pure rohu or catla seed. At that point of time, it is almost impossible for the farmers to recognize the differences. So after nursery rearing, poor survival and growth of these seeds become a bane for them. Even though they approach the fishery officials for law enforcement, they also feel helpless in the absence of a genuine identification method at this early stage.

An attempt has been made to review hybridization in carps (Cyprinidae), focusing on the advantages and disadvantages of human-mediated hybridization in cultured species of carps and detection methods of carp hybrids. Further, the development of a novel multiplex PCR-based approach for the identification of seeds of rohu and catla and their hybrids (parents and hybrids), using some molecular markers, and its field testing with hatcheries in four Indian states are described.

2. Hybridization: natural and human-mediated

The major purpose behind human-mediated hybridization is genetic improvement in cultured species. This is expectedly achieved through combining desirable traits of parental species, resulting in heterosis or hybrid vigor in the progeny. In fishes human-mediated hybridization is around 50% [23]. In natural hybridization reproductive barriers prevent introgression to happen. That is, chances of F1 hybrid mating with parental species are remote. On the other hand, in human-mediated hybridization in which the rate is higher and the number of escapees is more, the

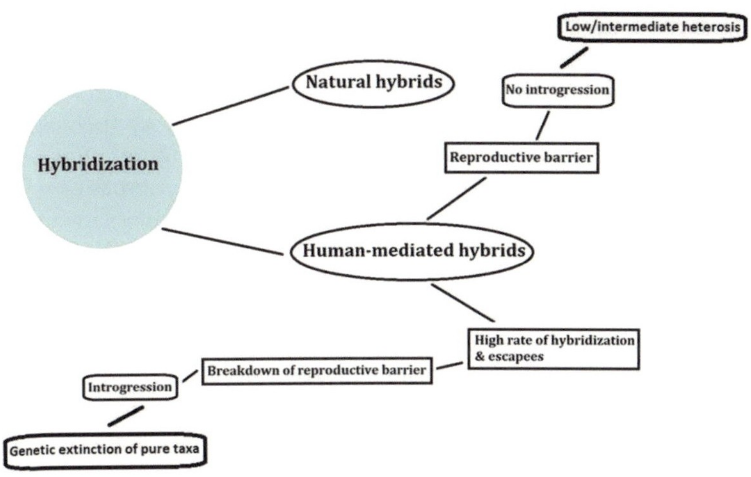

Figure 1.
Schematic representation of natural and human-mediated hybridization.

reproductive barriers are broken down, leading to introgression which may eventually lead to genetic extinction (**Figure 1**).

3. Hybridization in Cyprinidae

Hybridization is more widespread among members of Cyprinidae than any other groups of freshwater fishes [23]. Several permutation and combinations of interspecific and intergeneric crosses were carried out among Indian major carps (*Catla catla*, *Labeo rohita* and *Cirrhinus mrigala*) and exotic carps (*Hypophthalmichthys molitrix*, *Ctenopharyngodon idella*, *Aristichthys nobilis* and *Cyprinus carpio*) with the major objective to achieve hybrid vigor in economic traits [24–31]. The Indian major carp species are known to be able to hybridize, and hybrids are fertile and can be backcrossed to the parental species [32–34]. Hybridization has been shown to have a significant impact on production-related traits (notably growth), with some studies reporting a growth rate of F1 hybrids intermediate between that of the parent species [35, 36]. Hybrids are reported to have lower rates of growth than either of the parental species of Indian carps, other studies reporting growth rates lower than for either of the parental species [37]. Intergeneric hybrids between catla (*Catla catla*) and fimbriatus (*Labeo fimbriatus*) were produced by employing the technique of hypophysation and dry stripping. Detailed investigations on their embryonic and larval development, taxonomic characters, and aquaculture potential in terms of growth, feed utilization, body carcass composition, meat yield, etc. were carried out [28].

A catla-rohu hybrid produced by hypophysation was found to be intermediate in general appearance to the parent species. Gut content analysis revealed that the hybrid was mainly phytophagous in its diet. Growth rate was observed to be faster than in rohu. It matured within 3 years and was equally responsive to hypophysation [38]. In other cyprinid species, also such efforts of hybridization were carried out [23, 39].

3.1 Purposes and merits of human-mediated hybridization

It has long been recognized that hybridization can have a variety of evolutionary outcomes, including outcomes that maintain or increase diversity such as stable hybrid zones, the evolutionary rescue of small inbred populations, the origin and transfer of adaptations, the reinforcement of reproductive isolation, and the formation of new hybrid lineages [40, 41].

Hybridization in carps was being carried out to increase growth rate, transfer desirable traits between species, combine desirable traits of two species into a single group of fishes, reduce unwanted reproduction through production of sterile fish or mono-sex offspring, take advantage of sexual dimorphism, increase harvestability, increase environmental tolerances, and increase overall hardiness in culture conditions. Hybrids constitute a significant proportion of some countries' production for certain taxa, for example, hybrid striped bass in the USA, hybrid clarid catfish in Thailand, hybrid characids in Venezuela, and hybrid tilapia in Israel. Hybridization has been used in tandem with polyploidization to improve developmental stability in hybrid progeny [1]. Intergeneric hybrids between catla (*Catla catla*) and fimbriatus (*Labeo fimbriatus*) combined desirable qualities such as the small head of the fimbriatus and the deep body of the catla and exhibited heterosis in terms of meat yield with higher flesh content than either of the parents. Hence the hybrids appear to be of considerable importance to aquaculture [28]. It is also believed that hybrids of parental genotypes might be able to explore ecological niche unavailable to the latter. It can lead to adaptation through creation of novel genotypes and morphologies. Further, hybrid vigor or heterosis can also occur [39]. Interspecific hybrids are created purposely to enhance productivity of aquacultural strains [42]. Hybridization is done also to enhance recreational angling opportunities [43].

Evolutionary evidence that hybridization is a constructive process was mentioned, as (a) reproductive barriers, both pre- and post-zygotic, between described species appear incomplete among many fishes, (b) permanent transfer of genetic information apparently is possible even when hybrids and backcrosses are under negative selection, and (c) genetic exchange through introgressive events may have significant effects on the genetic composition of a species, and thereby, actually contribute to diversity within taxa [7]. Studies of interspecific hybridization between the Siberian sturgeon and Russian sturgeon showed that the hybrids had higher survival and growth than the purebreds under provided hatchery conditions. The hybrid crosses displayed positive average heterosis in most of the assessment times for growth and survival traits, and better fitness-related traits than purebreds, thus suggesting that interspecific hybridization provides a survival advantage to sturgeons during their evolutionary period [44].

4. Threats and concerns of human-mediated hybridization

The harmful effects of hybridization, with or without introgression, have led to the extinction of many populations and species in many plant and animal taxa. Hybridization is especially problematic for rare species that come into contact with other species that are more abundant [2]. Hybridization can decrease diversity through the breakdown of reproductive barriers, the merger of previously distinctive evolutionary lineages, and the extinction of populations or species. There are two main mechanisms by which hybridization can lead to extinction. If hybrid fitness is strongly reduced relative to that of parental individuals (i.e., outbreeding depression), and hybridization is common, population growth rates of one or both parental lineages may decline below replacement rates due to wasted reproductive effort, leading to extinction [41, 45]. Hybridization involving captive-bred individuals can have harmful consequences beyond the loss of genetic integrity [46]. In many cases, the stocked individuals differ genetically from the target population, which can result in outbreeding depression following hybridization [45].

Inadvertent hybridization and backcrossing can lead to unexpected and undesirable results in hybrid progeny, such as failure to produce sterile fish, loss of color pattern, and reduced viability. Uncontrolled and unintentional hybridization could undermine the performance of cultured stocks and restrict future use of the contaminated stocks as broodstock. The level of unintentional or accidental hybridization has important considerations for the conservation of aquatic biodiversity and will influence risk assessment on the use of hybrid fishes in aquaculture and fisheries [1]. Continued hybridization may eventually lead to a breakdown of species barriers, thereby compromising the genetic integrity of the species in the wild and leading to production losses in aquaculture [30]. Hybrid introgression in major carp species is very likely to have negative consequences, as a result of loss of distinct feeding strategies of the pure species, which are the basis of successful polyculture systems [46].

In Indian major carps, inadvertent production of hybrids out of mixed spawning of species has been well documented. Actually the farmers are on the lookout for "mixed seeds," meaning a certain proportion of catla, rohu, and mrigal along with other exotic carps for polyculture [47]. For the sake of time and economy, the hatchery producers keep broodfish of different species, particularly rohu and catla, in the same breeding pool, resulting in unintentional production of the hybrid seeds [34, 48]. The intergeneric hybrids are fertile, and they can breed (backcross) with parental species to produce introgressed F2 hybrids. The thoughtless and injudicious ways of fish breeding are likely to affect the "gene pools" of these prized food fishes badly [27].

Silver carp and bighead carp sometimes are hybridized inadvertently because of their similar appearance and because of shortage of "the correct" species at spawning time due to differences in maturation times between male and female carp. This hybridization often results in a fish that does not feed efficiently as its gill rakers are intermediate in shape between those of the silver carp that eats phytoplankton and those of the bighead carp that consumes zooplankton [1]. The rohu-catla reciprocal hybrids are reported to have limited economic value [27]. These hybrids are also reported to be more susceptible to parasitic infection than the parental species [49]. Hybridization between silver carp (*Hypophthalmichthys molitrix*) and bighead carp (*Aristichthys nobilis*) suggests further generations of hybridization or introgression between the species in hatcheries, with potentially damaging consequences for the integrity of these stocks and their performance in aquaculture [50]. Pecos pupfish (*Cyprinodon pecosensis*) is threatened with replacement by its hybrids with sheepshead minnow (*C. variegatus*) [12]. Continued hybridization between invasive bigheaded carps (*Hypophthalmichthys nobilis*) and silver carp (*Hypophthalmichthys molitrix*) has indicated reduced nutritional performance of their progeny [29].

5. Detection of hybrids

Accurate identification of hybrids is important not only for sustainable aquaculture development, guiding aquaculture domestication efforts, assessing aquaculture production, and identifying useful crosses, but also to allow for a better understanding of biodiversity issues. It would be unfortunate to experience a widespread loss of pure species in aquaculture as happened with tilapia as a result of widespread introduction and subsequent hybridization; it would be also a significant cause for concern if hybrid Thai catfish or the hybrid Venezuelan characids pose more of a threat to local species than the pure species [51].

Before 1966 only morphology-based methods were used to identify hybrids. Subsequently followed by morphology (45%), allozymes (35%), mtDNA (12%), nDNA (4%), and karyology (2%) were used till the late 1990s. Genetic markers and population genetic theory have provided powerful tools facilitating the description of hybridization events and serve as sources of evidence for factors underlying occurrence, direction, and extent of introgression between fish taxa [23]. VNTR minisatellite and microsatellite loci, SINE's, RAPD, AFLP, and ISSR assume dominance, whereby individuals are characterized by the presence or absence of amplification products of specific size. The number of alleles producing a product (one for heterozygotes and two for a homozygote) cannot be directly determined. Thus, the per-locus information context of dominant markers is less than for codominant loci. Mitochondrial DNA cannot be used alone to detect hybrids because of the marker's haploid and matrilineal mode of inheritance. However, mtDNA can be a powerful tool to establish directionality when used in conjunction with nuclear genetic markers [23]. Genetic markers, such as allozymes, mtDNA, and nuclear DNA, were used to confirm hybrid status and to determine directionality of the hybridization event [12]. Multiple markers were employed to determine if an Icelandic population of eels (*Anguilla anguilla*) included hybrid individuals from matings of parents originating from populations in North America and Europe [43].

Historically, meristic and morphometric measurements were the primary means of identifying naturally occurring hybrids. The introduction of allozyme electrophoresis provided a methodology whereby individuals of most species could be assayed for biochemical markers with a demonstrable heritable basis [52]. The use of mtDNA was first cited in the surveyed literature in analyses of fish hybridization in the mid-1980s [4]. In the late 1980s, nuclear DNA was started to be used for identifying hybridization process [53, 54].

Documentation of hybridization often has been based on meristic or morphological criteria that can be misleading when used as the sole source of inference, particularly for hybrid individuals beyond the F1 generation [55]. Morphology, allozymes, and mtDNA were used in the analysis of *Notropis chrysocephalus* and *N. cornutus* hybrid zones in Michigan drainages [56]. Genomic RFLP was used to detect hybrids of Indian major carps, and the results indicated that intergeneric hybridization did occur during "mixed spawning" of these carps and the hybridization frequency was appreciable, at about 10% [34]. Utilizing an integrated approach, which incorporates geometric morphometrics, life history, and molecular genetic analyses, the levels and processes of hybridization in two species of cyprinids were determined [39]. The extent of intergeneric hybridization in Indian major carps was studied using allozymes [30]. DNA fingerprinting using RAPD-ISSR assay was used to detect hybridization in Indian major carps [57].

6. PCR-based kit for detection of early hybrids of rohu-catla reciprocal crosses

6.1 Development

For the parental lineages, 50 individuals of *Labeo rohita* (rohu) and 50 individuals of *Catla catla* (catla) were genetically analyzed. Crosses performed by mating females of catla and males of rohu and vice versa resulted in the intergeneric hybrid (**Figure 2**). Twenty-four hybrid individuals were included in the genetic analysis. Spawns of reciprocal hybrids were collected for further genetic analysis. DNA was extracted from the fin clips of adults of parental species using standard phenol-chloroform method [58].

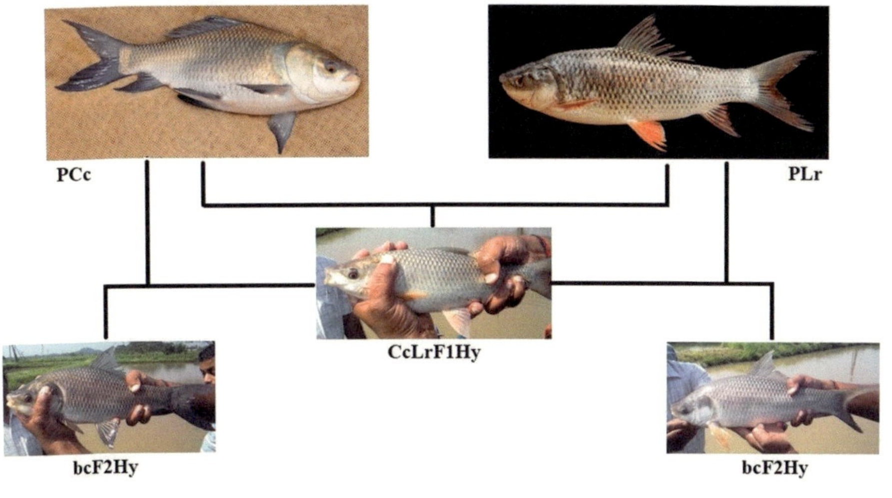

Figure 2.
Catla (PCc), rohu (PLr) parents, F1 hybrid (CcLrF1Hy), and F2 backcross hybrids (bcF2Hy).

Total genomic DNA from spawn was isolated using DNeasy blood and tissue kit, Qiagen. DNA quantity was determined against a molecular marker standard (λ-DNA 25 ng, Fermentas) by electrophoresis in a 0.8% agarose gel. β-actin sequences of carps available in GenBank were downloaded (Accession numbers: AF415205, M24113, GU338376, AY531753, AF415206) and aligned using Clustal W program implemented in the software Bioedit version 7.0.5.3 [59], and conserved primers for the amplification of a fragment size of ~ 1000 bp in Indian major carps and minor carps were done manually. Genomic DNA (~20–100 ng) from both species of IMCs was amplified in a 25 μl PCR volume containing 10 picomoles of each conserved primer, 2.5 mM of each dNTP, and 0.25 U of *Taq* polymerase with a thermal regime of 94°C (5 min), 35 cycles at 94°C (0.5 min), 60°C (0.5 min) and 1 min at 72°C (1 min) and final extension of 72°C (5 min). PCR products were purified using Qiagen PCR purification kit followed by bidirectional cycle sequencing on ABI 3100 PE automated capillary sequencer.

A total of 20 sequences of both the species (10 *Labeo rohita* and 10 *Catla catla*) were aligned using Clustal W program in Bioedit software. Species-specific reverse primers for both species were designed, taking the species-specific mutation into account. A touchdown PCR was carried out with a 25 μl PCR volume containing 10 picomoles of each species-specific reverse primers (one rohu and one catla) and 20 picomoles of universal forward primer, 2.5 mM of each dNTP, and 0.25 U of *Taq* polymerase with the PCR condition of 94°C (5 min), 2 cycles at 94°C (0.5 min), 68°C (0.5 min) and 1 min at 72°C (1 min), 2 cycles at 94°C (0.5 min), 66°C (0.5 min) and 1 min at 72°C (1 min), 2 cycles of at 94°C (0.5 min), 64°C (0.5 min) and 1 min at 72°C (1 min), 25 cycles at 94°C (0.5 min), 62°C (0.5 min) and 1 min at 72°C (1 min) and final extension of 72°C (5 min). The PCR products were checked in a 2% agarose gel.

Partial sequences of the nuclear β-actin gene amplified using a set of primers BAF (5′GTAGGCACGACATTGAATGGG3′) and BAR (5′AGACAAAGGAAGTCCCTCTGC3′) generated a total of 820 bp which revealed some differences in the nucleotide composition between *Labeo rohita* and *Catla catla*. Single-nucleotide polymorphism was found between the species which were used to design species-specific internal primers.

Two primers were designed specific for each species considering the polymorphic sites in the sequences. Both primers designed were in the reverse direction: primer BALRR (5′-CTTGAAAACTGTACAATCACGTTC-3′) was specific for *Labeo rohita*, and BACCR (5′-GCTAGCTAATAGACGTAATCATTTAG-3′) was specific for *Catla catla*. Amplification of these primers (BAF, BALRR, and BACCR) established different electrophoretic banding patterns when run in a 2% agarose gel. The result revealed one band at about 100 bp specific for *L. rohita* and another band at 300 bp specific for *C. catla*. In the rohu × catla hybrid, a heterozygote pattern was observed with two diagnostic bands, with each one inherited from one parental strain. Using these species diagnostic primers, a PCR-based rohu-catla hybrid identification kit was developed which has received provisional Indian patent number "343/KOL/2013 of 26.3.2013." For the validation of the developed kit, a total number of 685 samples from different places were screened which revealed that 54 out of them were hybrids (**Figure 3**).

6.2 Features of the kit

The use of a multiplex PCR marker in the present study revealed a distinct electrophoretic pattern between rohu and catla and their hybrid. The advantage of multiplex PCR is that it does not require the additional step of restricted enzyme digestion and can thus eliminate any post-PCR analyses as well as additional time and costs. On the other hand, there are limitations to the primer designs that should be taken into consideration. The primers should be specific and reliable in

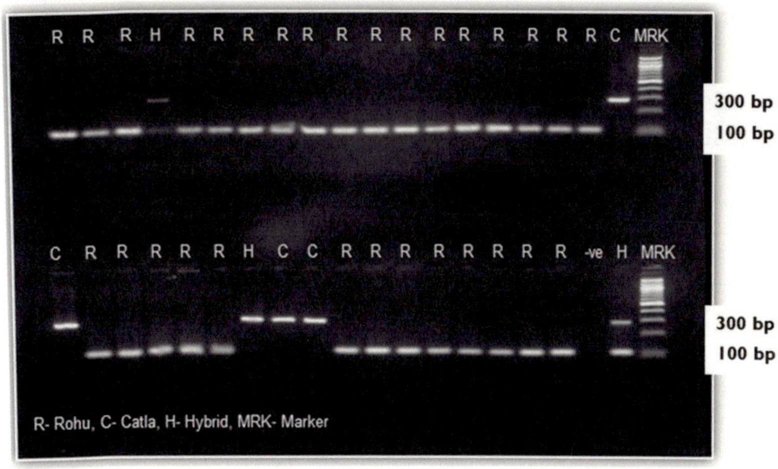

Figure 3.
PCR test of hatchery spawn samples (300 bp marker in catla, 100 bp marker in rohu, and both markers in hybrid).

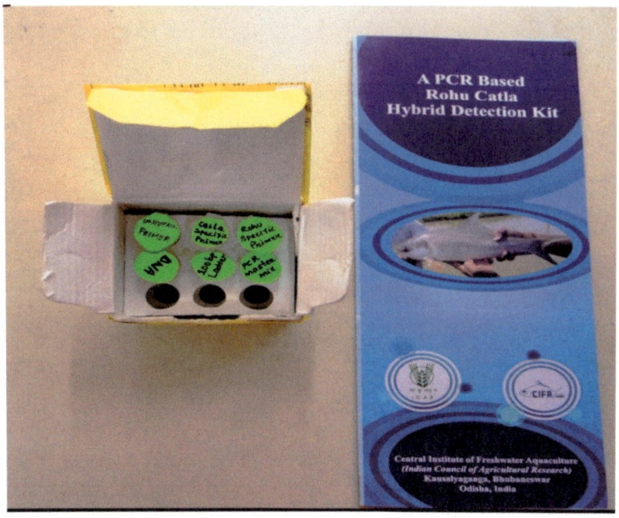

Figure 4.
Rohu-catla early hybrid identification kit.

binding. This study can serve as a basis for further study on the introgression of these hybrids with their parental species. Genetic monitoring of mixed spawning and unintended hybridization of Indian major carps in hatcheries can be carried out with the help of the kit (**Figure 4**).

Contents of the kit are mentioned below:

- Species-specific primers
- Universal primer
- dNTP mix (2.5 mM each)
- 10X Taq DNA buffer
- Taq DNA polymerase 3 U/μl

- Positive control DNA 100 ng
- Nuclease-free water
- 100 bp ladder

Advantages and utility of this kit are summarized below:

- PCR and agarose gel-based detection kit
- No sequencing required
- Takes only 4–6 h to get the results
- Highly sensitive and specific for rohu × catla hybrids and reciprocal crosses
- Useful for screening of hatcheries for genetic contamination
- Potential for seed certification hatchery accreditation
- An essential tool for government/private agencies to ensure purity of seed

6.3 Validation of primers by screening unknown samples from hatcheries

A total of 685 samples of different life stages (spawn and fry) from different hatcheries of four Indian states, Odisha, Gujarat, Bihar, and West Bengal (**Table 1**), were collected and tested using the species-specific designed primers with the same thermal cycler condition [60].

State	Details of hatchery	Samples (n)	Hybrid (n)	Hybrid (%)
Odisha	State Fisheries Farm, Kausalyaganga	150	6	4
Odisha	State Fisheries Farm, Bhadrak	30	0	0
Odisha	State Fisheries Farm, Balasore	30	2	7
Odisha	Balakati Private Hatchery	25	0	0
Odisha	Balakati Private Hatchery	25	3	12
Gujarat	Private Hatchery	45	23	51
Bihar	Shri Tripura Chaudhary, Matsya Farm, Vaishali	30	0	0
Bihar	Asha Fish Breeding Centre, Baheri Block, Darbhanga	30	3	10
Bihar	Kamla Fish Hatchery, Jhajarpur, Madhubani	30	2	7
Bihar	Koshi Fish Hatchery, Madhubani	30	5	17
Bihar	Ganga Fish Hatchery, Madhubani	30	4	13
Bihar	Yamuna Fish Hatchery, Madhubani	30	6	20
West Bengal	Naihati Market	200	137	68.5

Table 1.
Presence of hybrid seeds (spawn/hatchlings and fry) in hatchery populations in four Indian states (Odisha, Gujarat, Bihar and West Bengal), n = 685.

7. Conclusions

There have been numerous studies on hybridization of fishes, and certainly not all of the hybrids reported are contributing to commercial aquaculture production. Accurate identification of hybrids is important not only for sustainable aquaculture development, guiding aquaculture domestication efforts, assessing aquaculture production, and identifying useful crosses, but also to allow for a better understanding of biodiversity issues.

Intergeneric hybrids between the species rohu (*Labeo rohita*) and catla (*Catla catla*) are being produced in Indian carp hatcheries without any monitoring. The parental species belong to the more representative genus of the family Cyprinidae, which is an important freshwater fish group that is widespread throughout India, and are important fishery resources to specific communities. Some of the hatcheries are practicing multispecies breeding (mainly rohu and catla) in the same breeding pool and at the same time leading to interspecific hybridization. There is every chance of these hybrids escaping to natural waters which would lead to pollution of the genetic material in the wild, leading to non-availability of the pure strains of the carps in the future. There seems to be a misunderstanding regarding the culture of mixed species for composite fish culture with mixed breeding of carps by the hatchery managers and fish farmers. When the farmers are demanding mixed seeds (mixture of pure rohu and catla), the hatchery managers are mixing the spawners to produce hybrids.

Hybrid identification based on morphology, ecology, and behavior can be difficult and, most of the time, confusing and uncertain. Multiplex PCR strategy has proven to be an efficient methodology that could be quickly and inexpensively executed, which would allow diagnoses through simple PCR based upon single-nucleotide polymorphisms.

Morphological differences of hybrids are only minor and need close examination by experienced workers, and it is difficult to verify the genuine hybrids from interspecific hybridization. Allozymes were used to detect the genetic difference between the hybrids and their parental species; the use of allozyme loci failed to provide a sufficient genetic basis of hybrids, probably due to the less informative nature of allozyme loci (i.e., limited number of polymorphic loci available and low level of polymorphisms). Mutation at the DNA level that causes a replacement of a similarly charged amino acid may not be detected by allozyme electrophoresis, although allozymes represent actual gene products. Molecular techniques have been applied in the worldwide aquaculture, allowing for an adequate management of several cultivated species and providing a huge number of molecular markers that have been applied successfully for hybrid identification and detecting genetic introgression in fish. Nuclear molecular markers have supplied valuable information in the detection of hybridization events as well as the identification of reciprocal hybrids. Since mitochondrial DNA in animals has the characteristic of maternal inheritance [61], they are not suitable to detect hybrids. On the other hand, nuclear DNA serves as an efficient tool for hybrid identification.

The use of multiplex PCR marker revealed a distinct electrophoretic pattern between *Labeo rohita*, *Catla catla*, and their hybrid. The advantage of multiplex PCR is that it does not require the additional step of restriction enzyme digestion, which can eliminate any post-PCR analyses as well as additional time and costs. On the other hand, there can be limitations to the primer designs that should be taken into consideration, where these mainly reflect the ability of the primers to have good specificity and reliability in the application [62].

Finally based on the personal experience and inferences from other related studies, the following policy guidelines are recommended:

1. Natural hybrids should be eligible for protection.

2. Human-induced hybridization undesirable in majority of situations is a threat to conservation of parental species and low/intermediate heterosis. It causes extinction of pure taxa by replacement and genetic mixing. Hence it needs to be strictly regulated/eliminated.

3. Human-mediated hybrids shall be protected only in exceptional circumstances, such as when hybrids contain the only remaining genetic information from a taxon that has otherwise been lost by genetic mixing or when the circumstances of their origin are unclear.

4. In Indian major carps, mixed spawning of different species in hatcheries is to be stopped.

5. Good hatchery practices are paramount, and genetic monitoring of hatchery stocks on a regular basis is entailed to maintain the quality of fish seeds.

6. Unambiguous and rapid detection of hybrids at hatchery level is essential. The carp reciprocal early hybrid identification kit is useful for this purpose. Further, it is potentially useful tool for seed certification and hatchery accreditation.

Acknowledgements

I am thankful to my collaborators (Section 7) Mohanty M, Sahoo L, Das P, Das BK, Verma DK, and Routray P. This study was conducted when I was the Director of ICAR-CIFA, Bhubaneswar, during 2012–2017.

Conflict of interest

It is certified that there is absolutely no conflict of interest.

Author details

Jayasankar Pallipuram
ICAR-Central Marine Fisheries Research Institute (CMFRI), Ernakulam, Kerala, India

*Address all correspondence to: pjayasankar@yahoo.com

References

[1] Bartley DM, Rana K, Immink AJ. The use of inter-specific hybrids in aquaculture and fisheries. Reviews in Fish Biology and Fisheries. 2001;**10**:325-337

[2] Allendorf W, Leary RF, Spruell P, Wenburg JK. The problems with hybrids: Setting conservation guidelines. Trends in Ecology & Evolution. 2001;**16**(11):613-622

[3] Jhingran VG, Pullin RSV. A hatchery manual for the common, Chinese and Indian major carps. In: ICLARM Study Review. Vol. 11. Manila, the Philippines: ICLARM; 1988. p. 191

[4] Miao W. FMM/RAS/298: Strengthening capacities, policies and national action plans on prudent and responsible use of antimicrobials in fisheries. In: Final Workshop in Cooperation with AVA Singapore and INFOFISH; 12-14 December, Concorde Hotel, Singapore; 2017

[5] Sharma KR, Leung PS. Technical efficiency of carp production in India: A stochastic frontier production function analysis. Aquaculture Research. 2008;**31**(12):937-947. DOI: 10.1046/j.1365-2109.2000.00521.x

[6] Singh KD, Madan M, Rabbani MM, Sudhakaran AG, Pratheesh O, Ganesh T. Technical efficiency of freshwater aquaculture and its determinants in Tripura, India. Agricultural Economics Research Review. 2009;**22**:185-195

[7] Epifanio J, Nielsen J. The role of hybridization in the distribution, conservation and management of aquatic species. Reviews in Fish Biology and Fisheries. 2001;**10**:245-251

[8] Perry WL, Lodge DM, Feder JL. Importance of hybridization between indigenous and nonindigenous freshwater species: An overlooked threat to north American biodiversity. Systematic Zoology. 2002;**51**:255-275

[9] Senanan W, Kapuscinski AR, Na-Nakorn U, Miller LM. Genetic impacts of hybrid catfish farming (*Clarias macrocephalus* × *C. gariepinus*) on native catfish populations in Central Thailand. Aquaculture. 2004;**35**:167-184

[10] Na-Nakorn U, Kamonrat W, Ngamsiri T. Genetic diversity of walking catfish, *Clarias macrocephalus*, in Thailand and evidence of genetic introgression from introduced farmed *C. gariepinus*. Aquaculture. 2004;**40**:145-163

[11] Ryman N, Utter FM. Population Genetics & Fishery Management. Seattle: University of Washington Press; 1987

[12] Rosenfield JA, Nolasco S, Lindauer S, Sandoval C, Kodric-Brown A. The role of hybrid vigor in the replacement of Pecos Pupfish by its hybrids with Sheepshead Minnow. Conservation Biology. 2004;**18**(6):1589-1598

[13] Liu ZJ, Cordes JF. DNA marker technologies and their applications in aquaculture genetics. Aquaculture. 2004;**38**:1-37

[14] Hashimoto DT, Parise-Maltempi PP, Laudicina A, Bortolozzi J, Senhorini JA, Foresti F, et al. Repetitive DNA probe linked to sex chromosomes in hybrids between Neotropical fish *Leporinus macrocephalus* and Lepoirinus *elongatus* (Characiformes, Anostomidae). Cytogenetic and Genome Research. 2009;**124**:151-157

[15] Cocolin L, D'Agaro E, Manzano M, Lanari D, Comi G. Rapid PCR-RFLP method for the identification of marine fish fillets (Seabass, Seabream,

Umbrine, and Dentex). Journal of Food Science. 2000;**65**(8):1315-1317

[16] Magnussen JE, Pikitch EK, Clarke SC, Nicholson C, Hoelzel AR, Shivji MS. Genetic tracking of basking shark products in international trade. Animal Conservation. 2007;**10**:199-207

[17] Desai VR, Rao KJ. On the occurrence of natural hybrid of catla-rohu in Madhya Pradesh. Journal of Zoological Society of India. 1970;**22**(1/2):35-40

[18] Chaudhuri H. Fertility of hybrids of Indian carps and preliminary studies on the F_2 generation of carp hybrids. Journal of the Inland Fisheries Society of India. 1973;**5**:195-200

[19] Chondar SL. Systematic account of carp hybrid. l. *Labeo gonius* x *Labeo calbasu*. Journal of the Inland Fisheries Society of India. 1977;**9**:172-174

[20] Kowtal GV, Gupta SD. A note on the hybrid mrigal, *Cirrhinus mrigala* (Hamilton) x common carp *Cyprinus carpio* (Linnaeus) (Cyprinid). Aquaculture. 1984;**49**:179-183

[21] Ibrahim KH, Kowtal GV, Gupta SD. Contribution to hybridization among Indian carps. 1, Embroyonic and larval development in *Catla catla* (Hamilton) and *Hypophthalmichthys molitrix* (Valenciennes). Journal of the Inland Fisheries Society of India. 1980;**12**(2):69-73

[22] Little DC, Satapornvanit A, Edwards P. Freshwater fish seed quality in Asia. In: Edward P, Little DC, Demaine H, editors. Rural Aquaculture. UK: CABI Publishing; 2002. pp. 18-195

[23] Scribner KT, Page KS, Meredith L, Bartron ML. Hybridization in freshwater fishes: A review of case studies and cytonuclear methods of biological inference. Reviews in Fish Biology and Fisheries. 2001;**10**:293-323

[24] Chevassus B. Hybridization in fish. Aquaculture. 1983;**33**:245-262

[25] Khan HA, Kowtal GV. An account of naturally occurring and artificially produced cyprinid hybrids in India. NACA/WP/86/33 Network of Aquaculture Centres in Asia; Bangkok, Thailand; 1986

[26] Bhowmick RM, Jana RK, Gupta SD, Kowtal GV, Rout M. Studies on some aspects of biology and morphometry of the intergeneric hybrid, *Catla catla* Hamilton X *Labeo rohita* Hamilton produced by hypophysation. Aquaculture. 1981;**23**:367-371

[27] Tripathi SD. Three decades of research on carp hybridization in India. Proceedings of the Zoological Society. 1992;**45**(Suppl B):13-28

[28] Basavaraju V, Devaraj KV, Ayyar SP. Comparative growth of reciprocal carp hybrids between *Catla catla* and *Labeo fimbriatus*. Aquaculture. 1995;**129**:187-191

[29] Liss SA, Lamera JT, Sass GG, Suski CD. Physiological consequences of hybridization: Early generation backcrossing decreases performance in invasive bigheaded carps. Journal of Freshwater Ecology. 2016;**31**(4):543-554

[30] Simonsen V, Hansen MM, Mensberg K-ID, Sarder RI, Alam S. Widespread hybridization among species of Indian major carps in hatcheries, but not in the wild. Journal of Fish Biology. 2005;**67**:794-808

[31] Sultana S, Saifuddin Shah M, Ghosh AK, Bir J. Taxonomic study on the supposed hybrids of Indian major carps and Chinese carps in the local markets of Khulna, Jessore and Satkhira district, Bangladesh. Khulna University Studies. 2016;**13**(1):55-64

[32] Das P, Kumar D, Ghosh AK, Bhaumik U. Preliminary observation on the performance of catla-rohu hybrid

in carp culture. Food Farming and Agriculture. 1980;**12**:294-296

[33] Das P, Mishra A, Srivastava SK. Status on research in applied carp genetics and breeding in India. Journal of Aquaculture in the Tropics. 1996;**11**:307-317

[34] Padhi BK, Mandal RK. Inadvertent hybridization in a carp hatchery as detected by nuclear DNA RFLP. Journal of Fish Biology. 1997;**50**:906-909

[35] Chaudhuri H. Fish hybridization in Asia with special reference to India. Report to FAO/UNDP(TA). 1971;**2926**:151-159

[36] Natarajan AV, Desai VR, Mishra VN. On the natural occurrence of the intergeneric Catla-rohu hybrid in Rihand (Uttar Pradesh) with an account of its potential role in reservoir fisheries development in India. Journal of the Inland Fisheries Society of India. 1976;**8**:83-90

[37] Ibrahim KH. Contribution to hybridization among Indian carps: Studies on biological and morphometric features of some Indian carp hybrids [PhD thesis]. Bhubaneswar, India: Utkal University; 1977

[38] Bhowmick RM, Jana RK, Gupta SD, Kowtal GV, Rout M. Studies on some aspects of biology and morphometry of the intergeneric hybrid, *Catla catla* Hamilton X *Labio rohita* Hamilton produced by hypophysation. Aquaculture. 1981;**23**:367-371

[39] Hayden B, Pulcini D, Kelly-Quinn M, O'Grady M, Caffrey J, McGrath A, et al. Hybridization between two cyprinid fishes in a novel habitat: Genetics, morphology and life-history traits. BMC Evolutionary Biology. 2010;**10**:169

[40] Abbott RD, Albach S, Ansell JW, Arntzen SJE, Baird N, Bierne J, et al. Hybridization and speciation. Journal of Evolutionary Biology. 2013;**26**:229-246

[41] Todesco M, Pascual MA, Owens GL, Ostevik KL, Moyers BT, Hubner S, et al. Hybridization and extinction. Evolutionary Applications. 2016;**9**(7):892-908

[42] Lutz CG. What do you get when you cross. Aquaculture Magazine. 1997;March/April:84-90

[43] Avise JC, Van Den Avyle MJ. Genetic analysis of reproduction of hybrid white bass x striped bass in the Savannah River. Transactions of the American Fisheries Society. 1984;**113**:563-570

[44] Shivaramu S, Vuong DT, Havelka M, Šachlová H, Lebeda I, Kašpar V, et al. Influence of interspecific hybridization on fitness-related traits in Siberian sturgeon and Russian sturgeon. Czech Journal of Animal Science. 2019;**64**:78-88

[45] Muhlfeld CC, Kovach RP, Jones LA, Al-Chokhachy R, Boyer MC, Leary RF, et al. Invasive hybridization in a threatened species is accelerated by climate change. Nature Climate Change. 2014;**4**:620-624

[46] Mair GC. Genetics of broodstock management: Basic principles and practices. AARM Newsletter. 1999;**4**:4-6

[47] Jhingran VG. Fish and Fishery of India. New Delhi: Hindustan Publishing Corporation; 1991

[48] Padhi BK, Mandal RK. Improper fish breeding practices and their impact on aquaculture and fish biodiversity. Current Science. 1994;**66**:624-626

[49] Mishra BK. Observation on the susceptibility of the hybrid, *Labeo rohita#Catla catla* to *Argulus siamensis* (Wilson) infection in composite culture ponds. In: Proceedings of National Symposium in Freshwater Aquaculture.

Bhubaneswar, India: Central Institute of Freshwater Aquaculture; 1991. pp. 184-185

[50] Mia YM, Taggartb JB, Gilmourb AE, Gheyasb AA, Dasc TK, Kohinoora AHM, et al. Detection of hybridization between Chinese carp species (*Hypophthalmichthys molitrix* and *Aristichthys nobilis*) in hatchery broodstock in Bangladesh, using DNA microsatellite loci. Aquaculture. 2005;**247**:267-273

[51] RSV P, editor. Tilapia Genetic Resources for Aquaculture. Manila, Philippines: International Center for Living Aquatic Resources Management; 1988. p. 108

[52] Campton DE. Application of biochemical and molecular genetic markers to analysis of hybridization. In: Whitmore DH, editor. Electrophoretic and Isolectric Focusing Techniques in Fisheries Management. Ann Arbor, MI: CRC Press, Inc.; 1990. pp. 242-258

[53] Billington N, Herbert PDN, Ward RD. Evidence of introgressive hybridization in the genus *Stizostedion*: Interspecific transfer of mitochondrial DNA between sauger and walleye. Canadian Journal of Fisheries and Aquatic Sciences. 1988;**45**:2035-2041

[54] Hallerman EM, Beckman JS. DNA-level polymorphisms as a tool in fisheries science. Canadian Journal of Fisheries and Aquatic Sciences. 1988;**45**:1075-1087

[55] Neff NA, Smith GR. Multivariate analysis of hybrid fishes. Systematic Zoology. 1978;**28**:176-196

[56] Dowling TE, Smith GR, Brown WM. Reproductive isolation and introgression between *Notropis cornutus* and *Notropis chrysocephalus* (Family Cyprinidae): Comparison of morphology, allozymes, and mitochondrial DNA. Evolution. 1989;**43**(3):620-634

[57] Naveed R, Sultana S, Nawaz M, Ullah I, Al-Ghanim KA, Al-Thobaiti A, et al. Genetic variability in *Labeo rohita*, *Catla catla* and their hybrid (*Labeo rohita* ♂ × *Catla catla* ♀) populations employing randomly amplified polymorphic (RAPD)-inter simple sequence repeat (ISSR) assays. International Journal of Agricultural Biology. 2017;**19**:643-648. DOI: 10.17957/IJAB/15.0313

[58] Sambrook J, Fritsch EF, Maniatis T. Molecular Cloning: A Laboratory Manual. Cold Spring Harbor, New York: Cold Spring Harbor Laboratory Press; 1989

[59] Hall TA. BioEdit: A user-friendly biological sequence alignment editor and analysis program for Windows 95/98/NT. Nucleic Acids Symposium Series. 1999;**41**:95-98

[60] Mohanty M, Jayasankar P, Sahoo L, Das P, Das BK, Verma DK, et al. Early identification of hybrids of Indian major carp, *Labeo rohita* and *Catla catla* using nuclear marker-based multiplex PCR. Journal of Applied Ichthyology. 2016;**32**:963-965. DOI: 10.1111/jai.13111

[61] Moritz C, Dowling TE, Brown WM. Evolution of animal mitochondrial DNA: Relevance for population biology and systematic. Annual Review of Ecology and Systematics. 1987;**18**:269-292

[62] Pank M, Stanhope M, Natanson L, Kohler N, Shivji M. Rapid and simultaneous identification of body parts from the morphologically similar sharks *Carcharhinus obscurus* and *Carcharhinus plumbeus* (Carcharhinidae) using multiplex PCR. Marine Biotechnology. 2001;**3**:231-240

Chapter 3

Polymers and Microplastics: Implications on Our Environment and Sustainability

Vinod P. Sharma

Abstract

Polymeric are molecular structures that are built up primarily with multiple units of monomers, which may be natural or synthetic in nature. Amidst latest challenges with increased environmental awareness and pollution levels, there is increasing trend toward the production of plastics due to their extensive applications ranging from packaging to biomedical products. Plastics are versatile, indispensable, and cost-effective, require less energy to produce than alternative materials like metal or glass, and may be produced with different properties. They can be molded into different shapes, color, and functionality to serve the customers' expectations at affordable costs. Additives such as plasticizers, stabilizers, and colorants are added to monomers for specific properties as per requirement of manufacturing needs or guidelines. Although generally inert in nature, they may also pose health risks due to migration of few chemicals of environmental or health concerns in concentrations above than the permissible limits. Microplastics are now a great concern to environment and biodiversity. The extensive indiscriminate overuse of plastics, lack of adequate waste management practices, and casual community behavior toward their proper disposal pose a significant threat to the environment. Efforts are being taken globally for minimization of adverse implications, and concerted search are in progress for environment-friendly bio-based products with sustainability approaches.

Keywords: alternatives, applications, environment, microplastics, polymers, sustainable

1. Introduction

Synthetic materials of modern society are made from a range of organic polymers, viz., polyethylene (PE), polyvinyl chloride (PVC), polyethylene terephthalate (PET), polyethylene (PE), polypropylene (PP), polystyrene (PS), etc. Plastics are thus wide range of high molecular weight organic polymers obtained from the hydrocarbon and petroleum derivatives. We are experiencing increasing trend of production as well as consumption of plastics due to extensive applications in different sectors, viz., domestic, industrial, agricultural, health, etc. They may be molded into different shapes, color, and functionality. Additives added to monomers convert them into plastics with specific properties [1–17].

These may be available in various ranges of cost depending on ingredients, features, specifications, and applications. Based on production, economy, and demand, supply aspects' usage is dynamically changing. Plastics are versatile, indispensable, and cost-effective, and require less energy to produce than alternative materials like metal or glass, and may be produced with different properties for a more sustainable consumption in next few decades. Several advancement of human civilization has attributes and facilitation by the use of new polymeric products. As a consumer, one finds himself at crossroads at different stages of downsides of plastics and benefits [5–30]. There is a need to conduct SWOT analysis, determine gaps, and identify the opportunities so as to change the composition and practices of disposal for these invaluable polymers.

According to Water World report of 2016, the microscopic fibers of plastic may be released during each cycle of washing of clothes and ultimately reach to environment. According to WHO report of 2018, the microplastics may be in tap water or bottled water sources. It creates a concern to the scientific community, so as to potential preventive measures, and raise awareness at different levels.

Depending on characteristics, polymers are finding usage in packaging, spacecrafts, entertainment, toys industry, varied medical applications like disposable syringes, transfusion devices, intravenous bags, sterile or aseptic packaging for medical instruments, joint replacements, tissue engineering, intelligent, innovative packaging, etc.

Although generally inert in nature, they may also pose health risks. The biggest asset of durability is in fact the curse for the plastics in the recent scenario due to environmental consciousness. This is due to chemicals (EDCs), e.g., Bisphenol A or its analogs, di-(2-ethylhexyl) phthalate (DEI--IP), etc. [11–14]. The selection of the best packaging material depends on the cost viability, design that satisfies the competing needs in reference to the characteristics of the product, marketing considerations, environment and waste management issues, etc. In addition to these, it may also require a different analysis for individual item with consideration of properties of the packaging material, type of food to be stored, interactions, shelf life, environmental conditions, product end use, production, and distribution processes. The consumer wants to confirm that the products are capable of protecting the food ingredients and may not be damaged during transportation or storage till it is consumed under the specified period. Thus, roles of food packaging are very diverse and vary from perspective of regulators to stakeholders.

2. Degradability of polymers and anthropogenic interactions

Most of the polymers have the issue of degradability and creating marine pollution with degradation moieties which take couple of years. The extensive indiscriminate overuse of plastics, lack of adequate waste management practices, and casual community behavior toward their proper disposal pose a significant threat to the environment [2–10, 18–24]. Several animal species are on the verge of extinction, and thus biodiversity is slowly depleting. These are due to anthropogenic activities which comprise with life activities. Each species has specific and exclusive attributes. We are aware that the ecological integrity of the planet is under threat from exploitative activities of human beings. Moreover, the health of modern civilization, intellectual capabilities, and ethical and esthetic values are slowly fragmented with passage of time. Indiscriminate usage of plastics may be harmful to the environment, and one needs to curb excessive consumption. Let customers' be made aware for intuition within, rather than compel to buy bags with extra price to safe guard environment.

2.1 Degradation of plastics and effect at cellular/molecular level

Plastic breaks into smaller fragments which acquire properties that may increase the risk substantially. As per UN report, this may affect the ecological functions and impact on species exposed to toxic moieties. It is well known that the chemical effects are especially due to additives, viz., phthalates and Bisphenol A (BPA), poly brominated diphenyl ethers (PBDEs), which may migrate from plastic particles or under exposure conditions [11–17, 28–30]. The phthalates are well documented to disrupt the hormone system of the animal kingdom. Moreover, the nanoparticles may cause inflammation, cellular barriers, or cross the blood-brain barrier and the placenta. Cytological changes in gene expression and biochemical reactions may be triggered. More studies are in progress to understand the change in behavior or cognitive variations beside the implications on hepatic, nephrology, or reproductive systems. The genomic, transcriptomics, and metabolomic studies are yet to be established in this context.

2.2 Sustainability aspects: Environmental-friendly plastics

The environment lovers and academicians efforts toward sustainability need to be directed to make cost-effective, environment-friendly plastics. These are needed to safeguard the environment for future generations and instill a feeling of association with environment. The traditional plastics often create adverse effect on the environment due to indiscriminate disposal practices and over usage. Most of the drains and water bodies are choked with plastic or other associated waste. They may fill up landfills or end up as litter for several decades on land or in water and they are toxic to several animals. We may make the transformational change from being plastic user to environment-friendly bamboo toothbrushes, stainless-steel/glass water bottles, eco-friendly coffee tumblers, cloth pads/diapers, reusable shopping bags, compostable garbage bags, eco-friendly cutlery, etc.

2.3 Degradable polymers

Biodegradable plastics take 3–6 months to decompose fully depending on factors, viz., temperature, pressure, and the amount of moisture present. Biodegradable plastics have multiple applications—packaging, sutures etc. The economic, environmental, and health benefits are closely linked to health sector. The concept of materials coming from nature with environmental advantages of being biodegradable and/or bio-based is very attractive to the industry and to the consumers. In spite of increasing usage of bioplastics, they play a vital role in the fields of packaging, tapestry, buildings, defense, sports, agriculture, electronics, automotive, etc.

Biodegradable or bioplastics are generally perceived as the alternative solution for the waste management issues. Active, smart, or intelligent packaging is the need of the hour due to the fast changing consumer demands. Nowadays, there is another class of packaging which are known as responsive packaging. The waste from varied packaging forms a significant part of municipal solid waste and has caused increasing environmental concerns. There are requirements of strengthening various regulations so as to reduce the amounts generated.

3. Ecosystem and plastics

Several marine creatures die annually due to plastic entanglement and these are the ones found. Plastic does not decompose because very few organisms derive their energy from complex hydrocarbons such as oil depending on the plastic.

Plastic sheeting may break down under the UV light in sunlight within a couple of years. The aquatic ecosystem may include plants, animals, and micro-organisms present in ponds, rivers, beaches, and wetlands. Freshwater habitats are often classified by various factors, including temperature, light penetration, nutrients, and vegetation. Estuaries house flowers with the distinctive adaptation of having the ability to survive in contemporary and salty environments. Mangroves and Pickleweed are just few examples of estuarine plants. The fresh community is created from any of body of water that is made from fresh water like lakes, ponds, streams, and rivers [4, 8–10, 20–25]. They cover approximately 20% of the earth, and are in various locations spreading all over the globe. The Lentic ecosystem refers to stationary or relatively still water, from the Latin lentus, which means sluggish, e.g., lakes. Together, these two fields form the significant quantities of freshwater biosystems. Lentic systems are diverse, ranging from a small, temporal rainwater collection in a pool of few inches deep to Lake Baikal, which has a maximum depth of >1600 m. The major components of a freshwater ecosystem are producer, consumers such as zooplankton, diatoms, fish, turtles, and/or decomposers, viz., bacteria and fungi. Healthy, functioning fresh ecosystems give reliable and quality water flows upon which these basic human wants rely. Freshwater ecosystems, such as wetlands and rivers, also provide crucial regulating services, such as water purification, flood mitigation, and the treatment of human and industrial wastes.

The water quality plays a vital role in the process of leaching of materials. Leaching and permeation may occur in soft, inadequately buffered water from industries or varied processes of manufacturing units. The phthalic acid esters (PAEs), which are endocrine disruptors or hormonally active agents, are essentially added to provide flexibility and durability [4–8, 11–14, 18, 19, 28]. They have the ability to interfere with the endocrine system in the body of living organisms. With disposal of food packaging, cosmetic pouches, containers, bottles, toys, tubing's, transfusion bags, intravenous fluid bags, etc., used in medicals, they may reach to the ecosystem. PAEs are also associated with oxidative stress and alterations in cytokine expressions. The metabolites of phthalates in urine have been established to be associated with allergies and pulmonary implications in multiple studies or on public health. In vivo and in vitro studies are also in progress to understand the transgenerational and developmental effects.

4. Plastic recycling and issues

Plastic recycling is the process of reprocessing the waste into useful material and recovering the costly or useful metals. It is challenging under certain circumstances and needs good knowledge of interdisciplinary sciences. Recycling provides ample opportunities for reducing oil usage and carbon dioxide emissions. The advancements in recovering technologies and systems for the segregations, collection, sorting, and reprocessing of plastics and polymeric items are generating scope for innovations to divert the majority of plastic waste from landfills to generate energy or wealth from waste. Globally, life cycle analysis of few products reveals that substantial quantities of discarded end-of-life plastics are accumulating as debris in landfills or in natural habitats. Nowadays, recycling is one of the most dynamic areas for reducing negative impacts and represents creativity in few instances.

4.1 Plastic waste disposal

Wastes are inevitable in the progressive society, but we need to know how to minimize it or transform for something useful for society. Thermoset or thermoplastics waste is one of many types of wastes that take too long to decompose. Generally, a plastic item slowly decomposes in landfills depending on composition and environmental factors. The plastic bags we use in our everyday life take several years to decompose, while plastic bottles may take more than 10 decades, in few cases, depending on chemical composition. The Ministry of Environment, Forests and Climate Change in India has issued Gazette notification on plastics. The updated plastic waste management rules of the countries have laid down the process need to be automated and must take into account the ease of doing business for producers, recyclers, and manufacturers. The majority of plastic debris pollutes waters, accumulating in oceans. The social trends and behavior significantly affect occurrence of plastics in soil and thereby its transfer to other environmental compartments. The appropriate understandings of causative factors are essential for the proper design of preventive measures against pollution. It is said that *Ideonella sakaiensis* bacterium has the capability of plastic-eating proclivities. The waxworm caterpillars may breakdown plastic in a matter of hours, and mealworms possess gut microbes that may ingest polystyrene. Microorganisms, viz., bacteria and fungi through the production of active enzymes may be helpful in the biodegradability of polyethylene and pave a pathway for better future. This may be a natural process in the microbial world for carbon and energy source for their growth. This may be helpful in the recycling of materials in the natural ecosystem and utilized in the future. According to Japanese R&D on *J. sakaiensis* 201-F6 strains on polyethylene terephthalates, during 2016, various bacteria may work by secreting an enzyme (a type of protein that can speed up chemical reactions) known as PETase. This breaks few chemical bonds in PET, leaving smaller fragments that the bacteria may absorb using the carbon as a food source.

5. Plastics waste incineration: how safe and viable?

The combustion of plastic in the open air in presence of sufficient amount of oxygen leads to environmental pollution and release of toxic chemicals depending on the composition of basic materials and additives. The pulmonary disorders may be caused due to inhalation of toxic gases by humans and animals, and this may affect their health. The large quantities of non-segregated waste or residues remain even after the efforts of municipal authorities.

The hierarchy of waste management is termed as Lansink's ladder of reduce-reuse-recycle, and is being adopted by the European Union (EU) waste hierarchy. It may be helpful in the better utilization of resources, energy generation, and reduction of waste. The classified wastes subjected to incineration are, viz., municipal wastes, non-hazardous industrial wastes, hazardous wastes, sewage sludge, and clinical wastes. Incineration has few limitations, and the generations of toxic gases also need to be scientifically controlled. With recent developments, the ashes or unreacted leftovers/residues are used as road sub-base material and building sand and gravel alternatives. The concept of Rs are being applied worldwide, and Indian Standards are being updated as now we are living in open and free market scenario or economy [9, 24].

6. Innovations and Research and Development work in niche areas of plastics

The efforts are in progress for developing conductive composite which is biodegradable in nature [8, 16, 17, 29, 30]. Poly aniline (PANi) is an electro-active polymer with environmental stability and ease of synthesis via chemical or electrochemical mode from aniline or aniline hydrochloride monomer precursors. This has wide applications ranging from sensors to shielding, tissue engineering and antistatic coatings, etc.

Poly vinyl alcohol (PVA) nanofibrous matrix may be prepared by electro spinning an aqueous 10 wt% PVA solution. The PVA and chitosan scaffolds or nanofibers have been engineered for new applications in the tissue engineering and repair purposes. The water-resistance of the as-spun PVA nanofibrous matrix may be improved by physically cross linking the PVA nanofibers by heat treatment at 150°C for 10 min, which were found to be the optimal heat treatment conditions determined from chemical and morphological considerations. Due to its low water-resistance, some applications for PVA-based materials are limited (e.g., drug delivery systems and wound dressings). PVA mats containing tetracycline hydrochloride have been successfully developed by electro spinning. The water stability of the systems and cross linking of the PVA matrix may be induced by citric acid and heating in the range of 150–190°C for 3–5 min [20].

7. Futuristic plastics: plastics of next generation

The plastic materials of the next generation must be different scientifically and composition wise from the one which are being used currently. They may be blended with organic or inorganic materials to design tailor-made properties for specific usages [16, 30]. For drug delivery, biosensors, and tissue engineering, the biomaterials are being developed so that they disappear once the function has been fulfilled [16]. The bio-based plastics and composites made from readily available husk, molasses, tuber-based carbohydrates, corn, etc., are being used for developing new plastics with added specifications. Terms such as green, sustainable, or clean techniques are used for plastics from renewable sources with a special focus on the degradation of these polymers in natural environments. With regard to the former, we review innovations in feedstock development. The biodegradable end-product does not necessarily degrade once emitted to the environment as chemical additives used to make them fit for purpose will increase the longevity. The trend of upcoming market indicates new chemical entities made of nanocomposites, bio-based products, smart polymers, and degradable polymers of oxo or photodegradability features. Academicians are using green chemistry approaches to develop bioplastics, composites, innovative packaging models or to extract natural polymers, viz., cellulose, terpenes, waxes, etc., from vegetables and plant species for sustainability aspects.

8. Conclusion

The polymeric products are competitively less priced, strong, and durable with value added benefits to consumers. They may enhance the benefits for humankind. Plastics, generally, are not fast degradable. The plastics and several polymeric-based products may turn into microplastics with passage of time due to impact of environmental conditions which vary from region to region in the world. The aquatic bodies ranging from oceans to ponds may serve as sink for these degraded products or finer particulate matter. The marine birds and fishes are mostly affected with the ingestion of plastics. Organisms can also be seriously absorbed by floating plastic debris, or the contaminants may derive from plastic additives that are leached to the environment. Recent studies emphasize the important role of microplastics as they are easily ingestible by small organisms, such as plankton species, and form a pathway for contaminants to enter the food web. Contaminants leached from plastics tend to bioaccumulate in those organisms that absorb them, and chemical concentrations are often higher at higher trophic levels. This causes a threat to the basis of every food web and may have serious and far reaching effects, even on non-marine species such as polar bears and humans, who consume marinegrown food. Therefore, resolving the plastic debris problem is important to human kind for two reasons: we are the both creator and the victim of the plastic pollution problem.

The packaging design and innovation requires an integrated holistic approach in compliances to regulatory norms, so as to address vital issues and values throughout the entire chain of packaging and distribution [1, 5, 9, 24]. Solutions to the plastic waste issues may be attained through combined approaches of science and change in human behavior, especially in context to waste disposal practices. Such actions may include appropriate legislation against marine pollution by plastics, recycling and recovering processes must be encouraged, alternatives for degradable plastics must be motivated through innovative research and clean-up of debris must be a continual procedure using latest technologies. If the marine plastic pollution problem is to be resolved eventually, the Governments need continued support from the public and manufacturers/industrialists. Moreover, resolving this long-standing problem will require time, money, and energy from many individuals now living and those of future generations, if a safer and cleaner marine environment is to be achieved.

Acknowledgements

Author is thankful to Director, CSIR-Indian Institute of Toxicology Research, Lucknow for continuous encouragements. The interactions and deliberations on plastics in association with stakeholders were beneficial in writing this review.

Author details

Vinod P. Sharma
AcSIR, CSIR-Indian Institute of Toxicology Research, Lucknow, India

*Address all correspondence to: vpsharma@iitr.res.in; vpsitrc1@rediffmail.com

References

[1] American Society for Testing and Materials (ASTM). Plastics Standards and Terminology D 883-19 C, D 1600-18. Available from: http://www.astm.org/standards/plasticsstandards.html

[2] Brown AL. Freshwater Ecology. London: Heinimann Educational Books; 1987. 163p. ISBN: 0435606220

[3] Brönmark C, Hansson LA. The Biology of Lakes and Ponds. Oxford: Oxford University Press; 2005. 285p. ISBN: 0198516134

[4] Browne RA. Lakes as islands: Biogeographic distribution, turnover rates, and species composition in the lakes of central New York. Journal of Biogeography. 1981;**8**(1):75-83. DOI: 10.2307/2844594.JSTOR2844594

[5] Bureau of Indian Standards. Salient Guidelines viz IS 9845, 9833, 12709, on Plastics PCD 12 and 21; 2019

[6] Hillebrand H. On the generality of the latitudinal diversity gradient. American Naturalist. 2004;**163**(2):192-211. DOI: 10.1086/381004. PMID: 14970922

[7] Hillebrand H, Azovsky AI. Body size determines the strength of the latitudinal diversity gradient. Ecography. 2001;**24**(3):251-256. DOI: 10.1034/j.1600-0587.2001.240302.x

[8] Holmgren JI-I. Advantages of disposables: Safely, cost, convenience. Modern Hospital. 1974;**122**(1):35

[9] International Standardization Organisation (ISO). Guidelines on Plastics TC 61. Available from: http://www.iso.org/committee/49256.html

[10] Kale G, Kijchavengkul T, Auras R, Rubino M, Selke SE, Singh SP. Compostability of bioplastic packaging materials: An overview. Macromolecular Bioscience. 2007;7(3):255-277

[11] United Nations Environment Programme (UNEP). Converting Waste Plastics into a Resource, Assessment Guidelines. Osaka, Shiga: Division of Technology, Industry and Economics, International Environment Technology Centre; 2009

[12] US EPA Office of Water (4601M). Permeation and Leaching. 2002. Available from: http://www.epa.gov/sites/production/files/2015-09 documents/permeation and leaching.pdf

[13] US EPA Carbon Sequestrion in Landfills—US Environmental Protection Agency. Available from: http://epagov/climate change/vvaste/downloadsICFMemo CarbonSequestrioninlandfillspdf

[14] US EPA Teach Database on Phthalates Teach Chemical Summary. 2007. Available from: http://epa.gov/teach

[15] Sharma VP. Biobased composites: Future trends and environmental aspects. In: Smitthipong W, Chollakup R, Nardin M, editors. Biobased Composites for High Performance Materials: From Strategy to Industrial Application. Boca Raton, FL: CRC Press/Taylor and Francis Group; 2015. pp. 303-314. Available from: www.crcpress.com

[16] Sharma VP. Plastics of the future-innovations for improvement and sustainability with special relevance to biomedical applications. In: Depan D, editor. Biodegradable Polymeric Nanocomposites: Advances in Biomedical Applications. Boca Raton, FL: CRC Press/Taylor and Francis Group; 2016. pp. 221-230. ISBN: 97-1-422-6051-9

[17] Sharma VP, Singh RL. Degradable polymers and plastics of the next generation: Steps towards environmental sustainability, regulations and safety aspects. In: Principles and Applications of Environmental Biotechnology for a Sustainable Future Being. Springer; 2016

[18] North EJ, Halden RU. Plastics and environmental health: The road ahead. Reviews on Environmental Health. 2013;**28**(1):1-8. DOI: 10.1515/reveh-2012-0030. PMCID: PMC 3791860

[19] Fuhrmann G, Chandrawati R, Parmar PA, Keane TJ, Mavnard SA, BertazzoS, et al. Engineering extracellular vesicles with the tools of enzyme prodrug therapy. Advanced Materials. 2018;**30**(15):e1706616. DOI: 10.1002/adma.201706616. Epub Feb 23, 2018

[20] Kang YO, Yoon IS, Lee SY, Kim DD, Lee SJ, Park WII, et al. Chitosan-coated poly (vinyl alcohol) nanofibers for wound dressings. Journal of Biomedical Materials Research Part B: Applied Biomaterials. 2010;**92**(2):568-576. DOI: 10.1002/ jbm.b.31554

[21] Lambert S, Wagner M. Environmental performance of bio-based and biodegradable plastics: The road ahead. Chemical Society Reviews;**46**(22):6855-6871. DOI: 10.1039/c7csoo149e2017

[22] López-Córdoba A, Castro GR, Goyanes S. A simple green route to obtain poly(vinyl alcohol) electrospun mats with improved water stability for use as potential carriers of drugs. Materials Science and Engineering C: Materials for Biological Applications. 2016;**69**:726-732. DOI: 10.1016/j.msec.2016.070.21

[23] Narayan R. Biobased and Biodegradable Polymer Materials: Rationale, Drivers and Technology Exemplars. American Chemical Society Symposium Series. Vol. 939. ACS Publication; 2006. p. 282

[24] Organisation for Economic Cooperation and Development (OECD). Background Paper 1: Considerations and criteria for sustainable plastics from a chemicals perspective. ENV/JM/MONO, 5; 2019

[25] Coles R, McDowell D, Kirwan MJ. Food Packaging Technology. London: CRC Press; 2003. pp. 1-346

[26] Shah AA, Hasan F, Hameed A, Ahmed S. Biological degradation of plastics: A comprehensive review. Biotechnology Advances. 2008, 2008;**26**(3):246-265

[27] Sommer U, Gliwicz ZM, Lampert W, Duncan A. The PEG-model of seasonal succession of planktonic events in freshwaters. Archiv fiir Hydrobiologie. 1986;**106**:433-471

[28] Trasande L, Attrina TM, Bluestein J. Association between urinary bisphenol A concentration and obesity prevalence in children and adolescents. The Journal of the American Medical Association. 2012;**308**(11):u13-u21

[29] Sharma VP. Waste management and sanitation: Interrelationship, strategies and community participation Urban Panorama. Journal of Urban Governance and Management. Regional Centre for Urban and Environmental Studies [RCUES]. 2017;XVI(2): 121-132. S.no. 65. UGC approved Journal No: 48092. ISSN: 0975-8534

[30] Sharma VP. Polymeric membranes and composites—Innovations regulatory guidelines, developments for pollution control and environmental sustainability. In: Nasar A, editor. Smart Polymers and Composites, Material Research Forum. Vol. 21. Millersville PA, USA: MRF LLC; 2018. Available from: http://www.mrforum.com

Chapter 4

Bioavailability and Effects of Polystyrene Nanoparticles in *Hydra circumcincta*

Joëlle Auclair, Brian Quinn and François Gagné

Abstract

The release of nanoplastics (NPs) from the weathering and degradation of plastics in the environment is an important concern to aquatic ecosystems. The purpose of this study was to examine the bioavailability and toxicity of 50 and 100 nm fluorescently labeled polystyrene nanoplastics (NP) to the invertebrate *Hydra attenuata*. The hydrae were exposed to increasing concentrations of 50 and 100 nm NPs (1.25-80 mg/L) for 96 h at 20°C. A subgroup of hydra was depurated in media to determine the persistence of effects. The results revealed that the animals accumulated detectable amounts fluorescent NP and produced morphological changes at a threshold concentration between 5 and 10 mg/L. The hydrae were able to eliminate 76 and 78% of the 50 and 100 nm NPs, respectively. A characteristic tentacle detachment from the body was observed. Biochemical markers were also determined in exposed organisms and increased glutathione S-transferase (GST) activity, oxidative damage and neutral lipids levels that persisted after the 24 h. In conclusion, NPs are bioavailable to *Hydra*, produce morphological changes and increase oxidative stress and neutral lipids. The formation of neutral lipids could be the result of reduced food assimilation or a means for the elimination of NPs.

Keywords: nanoplastics, polystyrene, hydra, oxidative stress, neutral lipids

1. Introduction

The occurrence of plastics in the environment represents a major contamination problem, particularly for the aquatic environment. Indeed, it is estimated that over 8 million tons of plastic are released in oceans per year including micro- and nanoplastic (NP) fragments [1, 2]. Microplastic and NP are operationally defined as particles in the size range of 5 mm to 1 μm and 100–1 nm respectively, but plastics between 0.1 and 1 μm are considered by some authors to be NP [2, 3]. Given their small size, NPs can permeate cells and interact with biological macromolecules, possibly leading to unexpected long-term effects. This is further emphasized by the fact that at present NPs are difficult to measure in organisms due to a lack of bioanalytical methods. Recently, a method based on a fluorescent molecular rotor probe was proposed for NP bioanalysis in the attempt to address this bottleneck [4]. Among plastic polymers found in the marine environment, polystyrene, composed of vinylarene repeats, is the most abundant type accounting for up to 8% of total plastic production worldwide, with an annual production of more than 23 million tons per year [5, 6]. Plastics are subject to both abiotic and biotic weathering leading to degradation into the micro- and nanoscale, dramatically increasing the number of plastic-based nanomaterials in the environment.

As mentioned above, the small size of NPs allows them to be internalized by cells which could lead to severe cell damage such as apoptosis, protein denaturation/fibrillation and the formation of hydrophobic liquid crystals [7–9]. A number of laboratory studies have investigated the aquatic toxicity of NPs, primarily focusing on marine life, i.e. phytoplankton, zooplankton (crustacean, sea urchins) and bivalves (filter-feeders) [10–13]. NPs were shown to reduce algal growth, the development of sea urchin embryos and shrimp survival and produce immune-related effects in bivalves. These studies show that NPs could be harmful to marine organisms, but to date less information exists for freshwater invertebrates. For this reason, it is of value to include freshwater organisms when assessing the risks of nanomaterials given the potential for their behaviour to be influenced by surface water properties such as salinity, pH and organic matter [14, 15]. For example, in high-salinity environments, e.g. saltwater, low Zeta potential (surface charges), nanoparticles would tend to form aggregates, whereas aggregation would be less likely in low salt conditions [16].

The cnidarian *Hydra circumcincta Schulze, 1914*, is ubiquitously found in freshwater ecosystems and is often used as sensitive test species for toxicity assessments [17]. Hydra is considered an immortal organism which can regenerate for decades and is composed of a tube-like body (foot and digestive tube) with a head surrounded by six or more antennae to catch food [18]. The severity of toxic effects can be easily observed through characteristic morphological changes [19] (Figure 1S, supplementary data). Moreover, the impacts of chemicals or other stressors can be conveniently followed at multiple levels of biological organization such as behaviour (feeding activity), change in antennae and foot morphology, growth, regeneration, asexual reproduction and various biochemical markers (gene expression, proteomics, oxidative stress, genotoxicity, etc.) [20]. Hydrae are easy to maintain in culture and show unique regeneration activity. Recently, hydrae were used to investigate the effects of microplastics [21]. The study revealed that hydrae were able to ingest small microplastic fragments from facewash soaps that compromised feeding activity and displayed toxicity by morphological changes (antenna regression). The purpose of the current study was therefore to determine the bioavailability and toxicity of fluorescently labeled polystyrene NPs to hydra. Hydrae were exposed to 50 and 100 nm polystyrene NP and analysed for toxicity using morphological changes and biomarkers of energy metabolism (neutral lipids) and oxidative stress to relate the presence of NPs in hydra with toxicity responses.

2. Materials and methods

2.1 Preparation and exposure of NPs to hydra

Fluorescently labeled polystyrene nanoplastics were purchased from PolyScience (USA). Hydrae were maintained in 20-cm diameter crystallization bowls at 20°C in the following medium: 1 mM $CaCl_2$ and 0.5 mM TES (N-tris[hydroxymethyl]methyl-2-aminoethanesulfonic acid), pH 7.0, as previously described [17]. They were fed daily with *Artemia salina* brine shrimps. Exposure tests were conducted in 12-well microplates with three animals placed in each well (n = 9 organisms per treatment) in 4 mL of media. The NP suspension was pre-diluted at 0.5 g/L in Milli-Q water, and the final dilutions were prepared in hydra media. Controls consisted of the hydra media only. The exposure concentrations were 0, 1.25, 2.5, 5, 10, 20, 40 and 80 mg/L corresponding to 0.04, 0.079,

0.159, 0.318, 0.636, 1.27 and 2.54 × 10^{11} nanoparticles/L for 100 nm NP and 0.32, 0.64, 1.29, 2.57, 5.15, 1.03 and 2.06 × 10^{12} nanoparticles/L for 50 nm, respectively. Hence, on a mass basis, 50 nm NP contains eight times more particles than the 100 nm NP.

After a 96 h exposure at 20°C, changes in hydra morphology were determined based on the Wilby scale [18] from controls (normal-sized animals with long and slender tentacles) to sublethal effects (clubbed or reduced tentacles) and lethal effects (tulip or disintegrated stages). Representative examples for each morphological changes are represented in the supplementary data section (**Figure 1**). These morphological features were scored under a stereomicroscope at 6× magnification where the number of hydra with morphological changes is the endpoint used to calculate the 50% effect concentration (EC50) or 50% lethal concentration (LC50). A $ZnSO_4$ (LC50 of 1 mg/L; 0.7–1.3 mg/L 95% confidence interval) test was used to ensure reproducibility of culture conditions. At the end of the morphological assessment, the NP containing media was removed from the wells. The hydrae were washed in 1 mL of media, detached from the culture plate, suspended in groups of three organisms in 250 μL of a homogenization buffer and stored at −85°C for NP uptake and biomarker analyses. The homogenization buffer consisted of 50 mM NaCl containing 10 mM Hepes-NaOH, pH 7.4, 0.1 mM dithiothreitol and 0.1 μg/mL apoprotinin (protease inhibitor). A subset of hydrae at low concentrations (1.25, 2.5 and 5 mg/L) were allowed to depurate in clean media for 24 h before freezing to determine the elimination potential of the NPs and the toxicity potential.

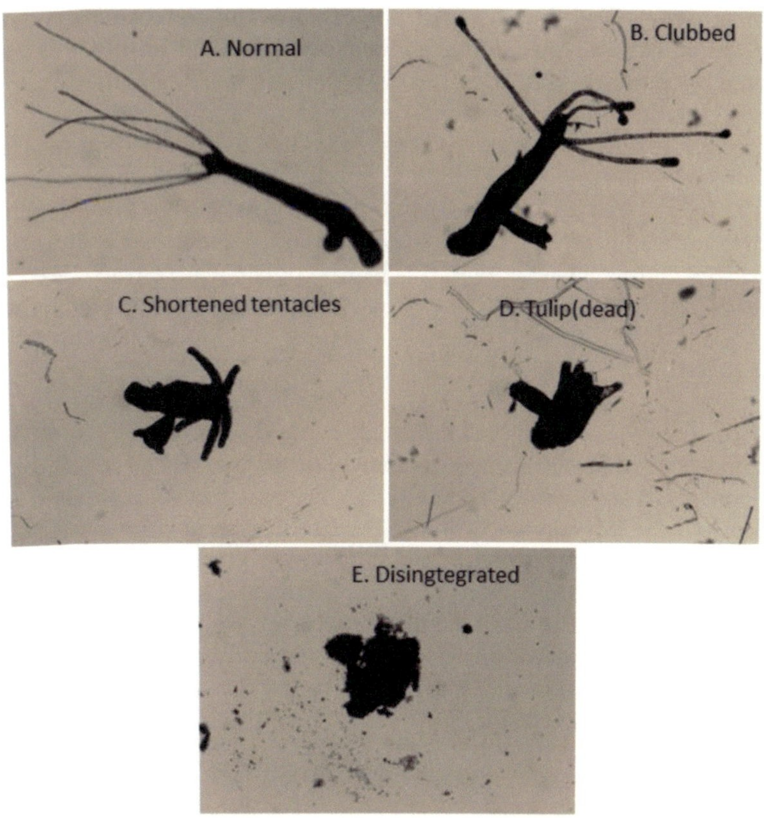

Figure 1.
Morphological stages Hydra attenuate after exposure to REEs for 96h (magnification x40): healthy polyp (A); clubbed tentacles (B); shortened tentacles (C); tulip stages (D) and disntegrated stages (E).

2.2 NP uptake and sublethal effect assessments

The organisms were thawed on ice for 15 min in the homogenization buffer and homogenized using a Teflon pestle handheld tissue homogenizer. Total protein content was determined using the Coomassie Brilliant Blue method in clear 96-well microplates [22]. Serum bovine albumin was used for calibration. The uptake of fluorescently labeled NP was determined in the homogenates using fluorescence spectrometry. A 100 μL volume of the homogenate was placed in 96-well dark microplates, and fluorescence levels were determined at 485 nm excitation and 520 nm emission (Synergy 4, BioTek Instruments, USA). Fluorescence concentration was determined using the standard addition methodology with freshly prepared fluorescence standards. The data were expressed as μm fluorescence/mg proteins. The activity of glutathione S-transferase (GST) was determined by a miniaturized spectrophotometric methodology in 384-well microplates (50 μL assay volume) [23]. The homogenate was mixed with 0.1 mM of reduced glutathione and 2,4-dichloronitrobenzene (DCNB) in 50 mM NaCl containing 10 mM Hepes-NaOH, pH 7.4, and incubated at 30°C for 30 min. The increase in absorbance at 340 nm was determined using a microplate reader every 5 min. The data were expressed as the increase in absorbance at 340 nm/min/mg of proteins. Lipid peroxidation (LPO) was determined in hydra homogenates using a miniaturized version of the thiobarbituric acid method [24]. A volume of 10 μL of the homogenate was mixed with 20 μL of 10% trichloroacetic acid containing 1 mM $FeSO_4$ and 20 μL of 0.7% thiobarbituric acid in 384-well dark microplates. The mixture was heated at 75°C for 10 min and cooled at room temperature. Fluorescence readings were taken at 540 nm excitation and 600 nm emission using standard solutions of tetramethoxypropane (stabilized form of malonaldehyde) for calibration in the blank media (homogenization buffer only). Results were expressed as μg thiobarbituric acid reactants (TBARS)/mg total protein in the homogenate. Finally, neutral lipids (triglycerides) in homogenates were determined using an AdipoRed fluorescent reagent (Lonza; Walkersville, MD, USA). A volume of 5 μL of AdipoRed reagent was added to 10 μL homogenate and 35 uL Milli-Q water in a black 384-well microplate. After 10 min under constant agitation, fluorescence was measured at 485 nm excitation and 535 nm emission (Synergy 4, BioTek microplate reader, USA). Data were expressed as relative fluorescence units (RFU)/mg proteins in the homogenate.

2.3 Data analysis

The study design examines the bioavailability and toxicity of 50 and 100 nm NPs in hydra exposed to seven NP concentrations. The 96-h toxicity (50% lethal concentration or LC50) was determined using the Spearman-Karber method [25] and the 96-h sublethal effect concentration (50% effect concentration or EC50) using the log-logit method. Data normality and homogeneity of variance were verified using the Shapiro–Wilk and Bartlett tests, respectively. The data were analyzed using analysis of variance and critical differences between the controls, and exposure concentrations were determined by least square difference (LSD) test. The trends between the data were also analysed using Pearson-moment correlation and hierarchical tree tests. All tests were performed using the SYSTAT software (version 13). Significance was set at $\alpha < 0.05$.

3. Result and discussion

The stock solutions of the fluorescently labeled polystyrene NPs contained 6.7×10^{14} and 8.3×10^{13} particles/mL for the 50 and 100 nm diameter NPs, respectively.

Hence, on a mass basis, there were eight times more 50 nm NPs than 100 nm NPs. The levels of fluorescently labeled NP were quantified in hydra exposed to increasing concentrations of NPs after 96 h and following a 24 h depuration for the low-exposure concentrations (1.25–5 mg/L) only (**Figure 2**). The hydra contained NPs in a concentration-dependent manner for concentrations >2.5 mg/L and >5 mg/L for 50 and 100 nm diameter NPs, respectively. Based on the ratio of the fluorescence label on each NP sizes, the hydra accumulated in the order of 6.448×10^7 NP per hydra exposed to 80 mg/L 50 nm NP and 2.605×10^8 NP per hydra exposed to 80 mg/L 100 nm NP. This suggests that the 100 nm NP were more bioavailable to hydra than the 50 nm NP. It is expected from the low salt content in the incubation media that the hydrodynamic behaviour of the NP would not overly change and minimal aggregation would occur [26]. The toxicity was determined in hydra based on morphological changes (**Table 1**). Although the lethal toxicity was not detected at concentrations below 80 mg/L based on the formation of tulip and disintegrated stages, the occurrence of the last morphological changes before reaching irreversible/lethal stage (short tube with receded tentacles) occurred at EC50 values of 16.2 and 8.7 mg/L for the 50 and 100 nm NPs, respectively. Based on the confidence intervals, the 100 nm NP was more toxic than the 50 nm NPs. A peculiar morphological response was observed during the depuration step in hydra exposed to 1.25–5 mg/L NPs for 24 h in clean media. Many hydrae lost some of their antenna during this period, and they appeared more fragile during removal for tissue processing as evidenced by fragmentation of the tentacles and body. Interestingly, the levels of NPs were still significantly higher albeit at lower concentrations than controls at 5 mg/L NPs for both sizes. It was estimated that it remained 3.34×10^3 particles per hydra for the 100 nm NPs and 2.6×10^5 particles per hydra for the 50 nm NPs. This represents an elimination potential of 79 and 76% for the 100 and 50 nm NPs, respectively. Hence, although the 100 nm NP were more bioavailable, the 50 nm NP were more slowly eliminated in hydra indicating a size-dependent effect. In a recent study, the 48 h lethal concentration (LC50) of a 75 nm diameter polystyrene NP was 77 mg/L in *Daphnia pulex* [27] which is in the same order with the observed toxicity at 8.7 and 16.2 mg/L for 100 and 50 nm NPs, respectively. This study also found that stress defense genes for superoxide dismutase and GST activity followed a biphasic response, i.e. were first induced and then inhibited in daphnids. The biphasic responses in GST activity and LPO levels in hydra exposed to the 50 nm NPs occurred at concentrations approaching manifestation of morphological changes in hydra.

Oxidative damage and xenobiotic conjugation activity were determined in hydra exposed to the NPs by following changes in GST activity and LPO, respectively (**Figure 3**). In hydra exposed to 50 nm NP, GST activity followed a biphasic response where a significant (LSD test, $p < 0.05$) decrease in activity was observed at low concentrations (2.5 mg/L) followed by an increase at higher concentrations compared to 2.5 mg/L. A decrease in GST activity was observed after the 24 h depuration period. In respect to LPO, an inverse trend was observed with a significant increase in LPO was found at low concentration of 50 nm NPs (5 mg/L) with a significant decrease at 10 mg/L. In the attempt to determine whether the observed changes in GST activity was related to LPO levels, i.e. involved in reactive oxygen species scavenging, an analysis of covariance was performed with LPO as the covariate and the exposure concentration as the main factor. The analysis revealed that GST activity was significantly related with exposure concentration and LPO had no significant effect on its activity. However, during the depuration period, both GST and LPO activity were increased which suggests a delayed stress response. Based on analysis of covariance, LPO did not influence GST activity during the depuration period as well which suggests that GST activity was more involved in

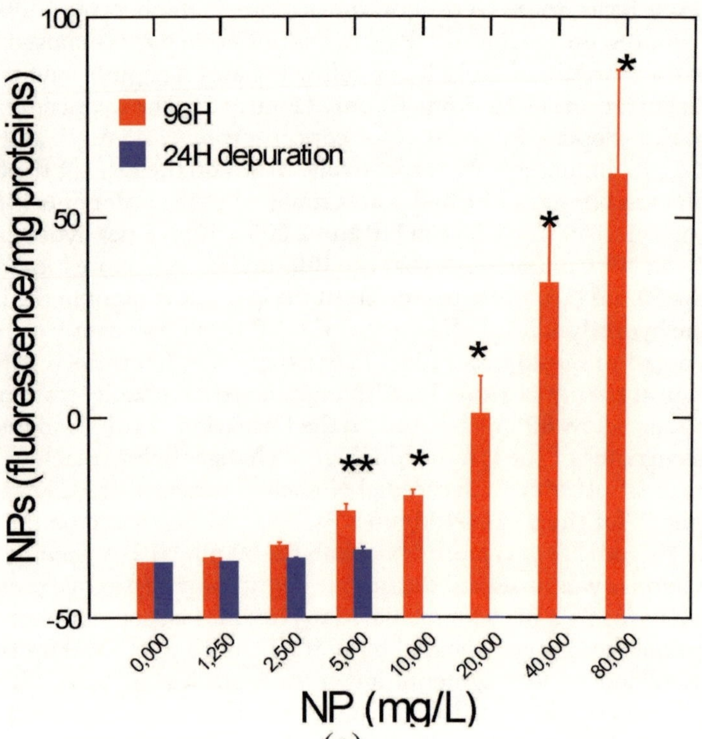

(a)

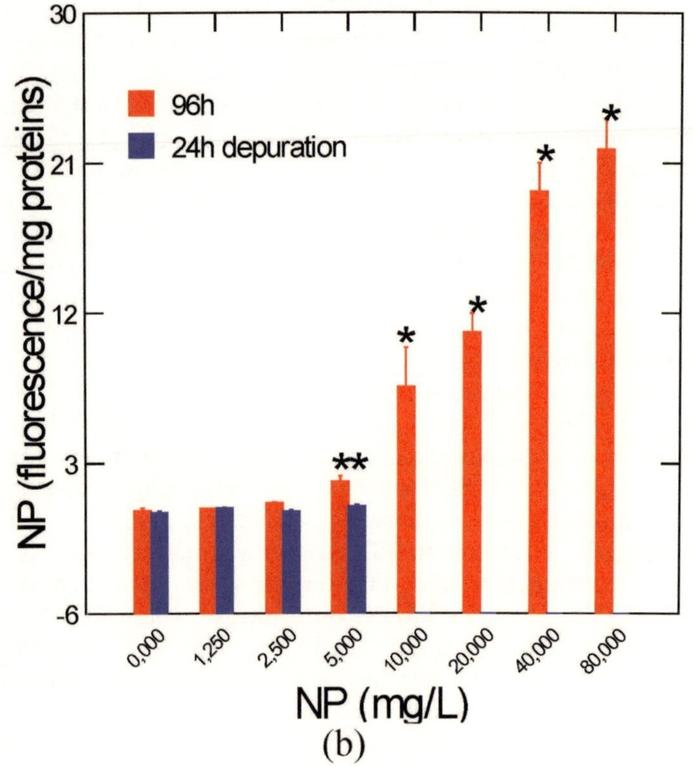

(b)

Figure 2.
Bioavailability of NP in Hydra attenuata *exposed to increasing concentrations of 50 nm (A) and 100 nm (B) polystyrene nanoplastics for 96 h at 20°C. A subsample of organisms were placed in clean media for 24 h (depuration). *indicates significance from controls (0) at $p < 0.05$. ** = $p < 0.01$.*

NP (nm)	LC50 in mg/L (95% confidence)	EC50 in mg/L (95% confidence)	Antennae loss in mg/L (amputation) during depuration
50	>80	8.7 (7–10)	1.25–5
100	>80	16.2 (11–26)	1.25–5

Time required for the manifestation of sublethal and lethal effects.

Table 1.
Toxicity data of 50 and 100 nm polystyrene NPs.

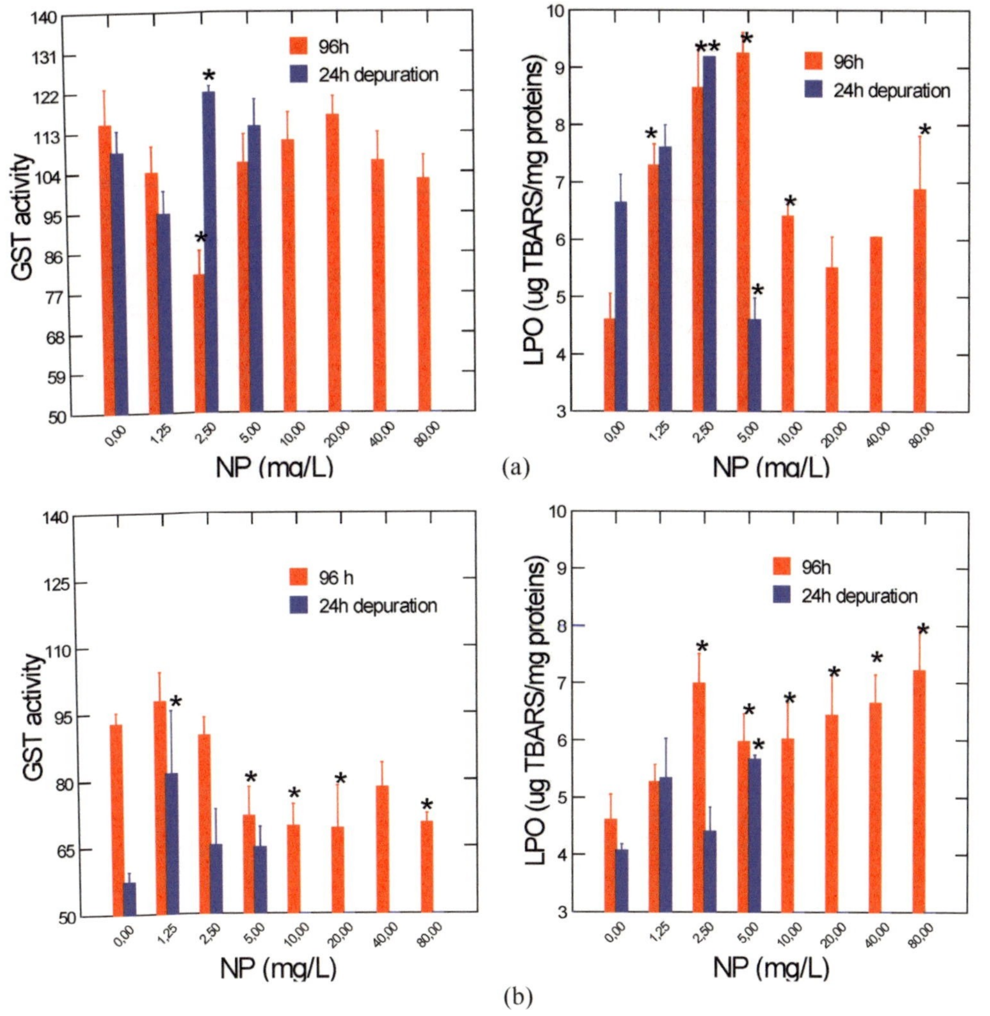

Figure 3.
Oxidative damage in Hydra attenuata *exposed to polystyrene NP. Hydrae were exposed to increasing concentration of 50 nm (A) and 100 nm (B) polystyrene NP for 96 h t 20°C. A subsample of hydrae were placed in clean media for 24 h (depuration). *indicates significance from the controls at p < 0.05.*

xenobiotic conjugation activity than in oxidative stress. In the case of the 100 nm NPs, overall GST activity was significantly decreased (ANOVA $p < 0.05$) with the concentrations used. A significant reduction in GST activity was observed at all concentrations >5 mg/L after 96 h exposure. LPO levels were significantly higher (ANOVA $p > 0.05$) than controls at concentrations >2.5 mg/L following the 96 h exposure. Since LPO results generally followed the same trend as the GST results, we performed an analysis of covariance of GST activity with LPO levels as the covariate to determine the contribution of oxidative stress on GST activity. The analysis revealed that both the exposure concentration and LPO levels significantly influence GST activity but the activity was more strongly associated with exposure concentration than LPO levels. This suggests that GST activity was not solely involved in handling oxidative stress but in conjugation activity as well. During the 24 h depuration period, both GST activity and LPO levels were decreased following exposure to 100 nm NPs suggesting a post-exposure stress reaction for the elimination of NPs. Exposure of hydra to NPs revealed that oxidative stress was readily increased by both diameter sizes. Moreover, LPO levels were related with neutral lipids in hydra suggesting that uncharged polystyrene NPs could lead to oxidative lipid damage in hydra. Oxidative stress was also found in the gut of *Daphnia magna* exposed to 100 nm NP [25]. Increased in LPO was also observed in *Artemia salina* shrimps exposed to 50 mg/L of polystyrene nanospheres [28]. Moreover, a reduction in GSH levels was observed which is consistent with the observed decreased in GST activity (depleted GSH?) in hydra exposed to the 100 nm; albeit a transient decrease in the activity was found with the 50 nm at low concentrations.

The levels of neutral lipids were also determined in hydra exposed to NPs (**Figure 4**). For the 50 nm NPs, the neutral lipids levels significantly decreased at low concentrations (1.25–5 mg/L) followed by an increase at higher concentrations where morphological changes appear (with the exception of 20 mg/L). Neutral lipids were highly correlated ($r = 0.81$, $p < 0.001$) with NP levels in hydra. After the 24 h depuration period, lipid levels remained significantly elevated at 5 mg/L NP. For the 100 nm NP, the lipid levels significantly increased at concentrations ≥5 mg/L following 96 h exposure. Neutral lipids were strongly correlated ($r = 0.99$, $p < 0.001$) with NPs, GST activity ($r = 0.48$, $p < 0.05$), and LPO ($r = 0.49$, $p = 0.01$) in hydra. After the 24 h depuration period, the neutral lipid levels were still significantly elevated at 5 mg/L. There was no significant correlation between neutral lipid levels with either GST activity or LPO levels with the 50 and 100 nm NPs. In the attempt, to gain some overview in the responses of hydra exposed to NPs, a cluster tree analysis was performed (**Figure 5**). The analysis revealed two response patterns. The first response pattern was that NP levels in hydrae were closely associated with neutral lipids for NP of both diameter sizes. The second pattern was that GST activity was closely related with LPO with the 100 nm NPs only. The involvement of neutral lipids with NP exposure and accumulation suggests altered metabolisms leading to increased neutral lipid mobilization. In the nematode *C. elegans* exposed to polystyrene NPs, there was increased gene expression of FOXO transcription factor suggesting that the insulin signaling pathway was affected [29]. Moreover, downstream genes in oxidative stress (superoxide dismutase and metallothionein) were involved which supports the hypothesis that polystyrene NP induces oxidative stress in invertebrates. It was found that a mutation of the FOXO gene increased the toxicity of NP which suggests that metabolic energy was required for protection against nanotoxicity of plastics. The increase in neutral lipid levels in hydra exposed to the NPs could be involved in detoxification/protection mechanisms to remove hydrophobic NP from cells. It was suggested that altered lipid metabolism could be obtained from the long-term exposure to both microplastics and NPs [30]. It is possible that the physical obstruction of the digestive track by NPs could lead to decreased food assimilation which could lead to increased mobilization of lipid energy reserves. Another explanation to increased neutral lipids in hydra is that it could represent a means of elimination of neutral hydrophobic nanoparticles composed on polystyrene (vinylarene polymer). Indeed, lipids could form a corona on theses NPs which could favor elimination by fusion to autophagosome and lysosome

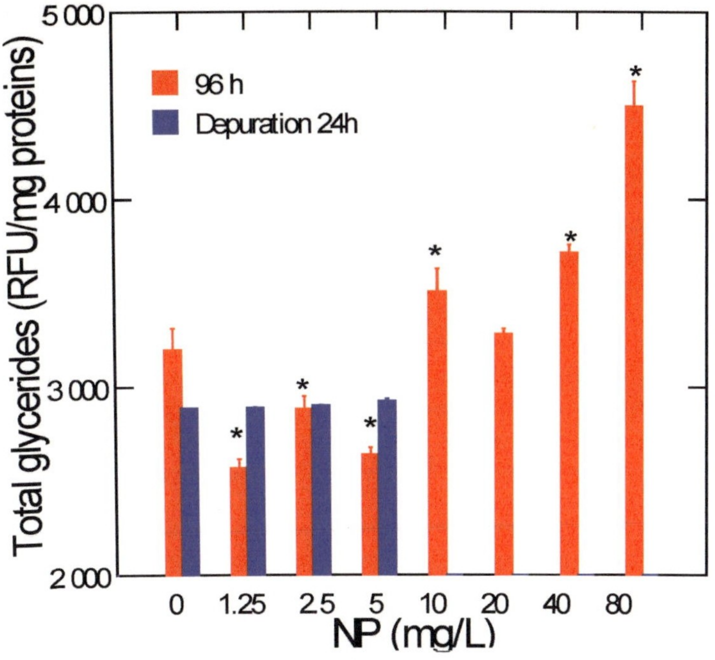

(a)

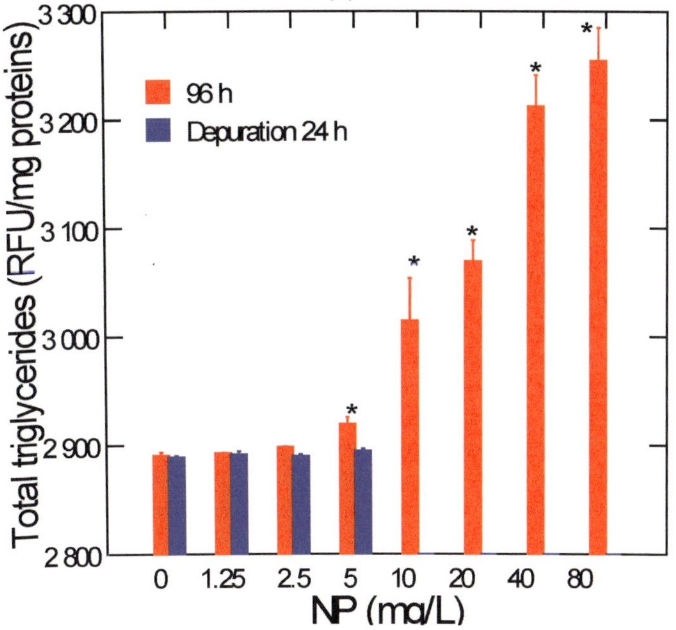

(b)

Figure 4.
Neutral lipid content in Hydra attenuata *exposed to NPs. The levels of neutral lipids were determined in hydra exposed to 50 nm (A) and 100 nm (B) polystyrene NP for 96 h t 20°C. *indicates significance (p < 0.05) relative to controls.*

organelles. This is in agreement with a previous study showing that the addition of lipids and proteins to 50 nm NP suspension in the presence of a membrane-selective probe leads to anisotropic changes [9]. Moreover, fluorescent polarization induced by NPs was significantly correlated with lipid contents ($r = 0.65$, $p < 0.001$) of the subcellular fraction of digestive glands of freshwater mussels. In conclusion,

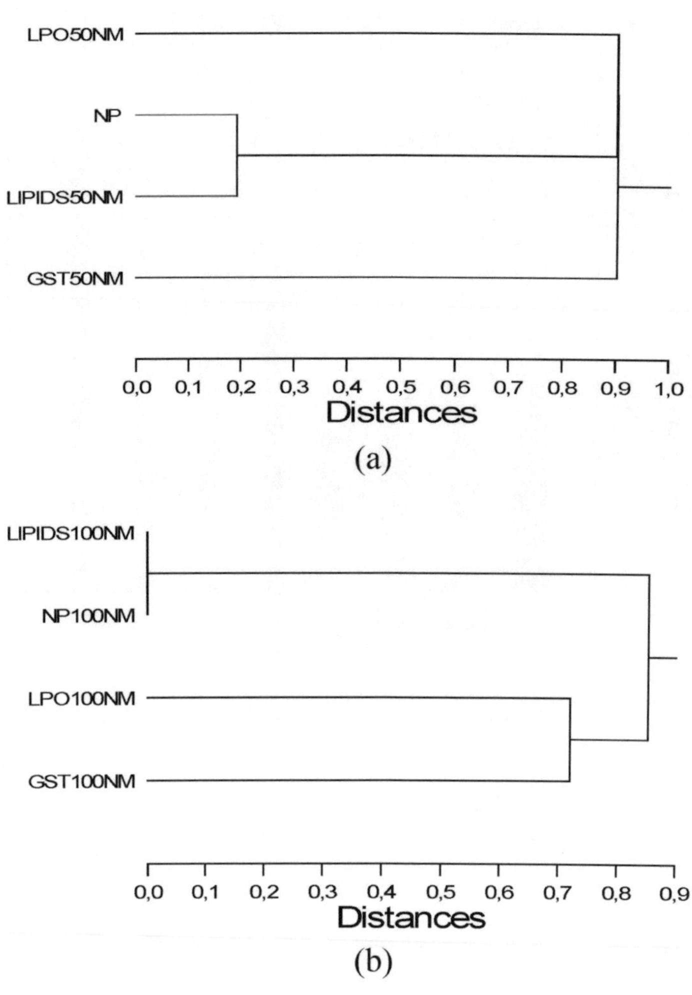

Figure 5.
Cluster analyses of biomarker data in Hydra attenuata *exposed to NPs. The biomarker responses were analysed using cluster analysis with the Euclidian methodology for distances (horizontal axis). The analysis for 50 nm (A) and 100 nm (B) NPs is shown.*

exposure of NPs to hydra led to a significant increase in NP accumulation which was still detected after 24 h depuration period in clean media. Hydrae exposed to NPs were more fragile and lost some antennae during depuration. The exposure also led to increased levels in neutral lipids, oxidative stress and xenobiotic conjugating activity which was also detected after 24 h depuration. NPs are bioavailable to hydra and produce morphological changes and sublethal effects such as oxidative stress.

Acknowledgements

The authors thank Pascale Gagnon from the Quebec Bioanalytical Laboratory of Environment and Climate Change Canada for performing the hydra exposure tests.

Author details

Joëlle Auclair[1], Brian Quinn[2] and François Gagné[1]*

1 Aquatic Contaminants Research Division, Environment and Climate Change Canada, Québec, Canada

2 School of Health and Life Sciences, University of the West of Scotland, Scotland, United Kingdom

*Address all correspondence to: francois.gagne@canada.ca

References

[1] Jambeck JR, Geyer R, Wilcox C, Siegler TR, Perryman M, Andrady A, et al. Marine pollution. Plastic waste inputs from land into the ocean. Science. 2015;**347**:768-771

[2] Mattsson K, Hansson LA, Cedervall T. Nano-plastics in the aquatic environment. Environmental Science: Processes and Impacts. 2015;**10**:1712-1721

[3] Gigault J, Halle A, Beaudrimont M, Pascal P-Y, Gauffre F, Phi T-L, et al. Current opinion: What is a nanoplastic? Environmental Pollution. 2018;**235**:1030-1034

[4] Gagné F. Detection of polystyrene nanoplastics in biological tissues with a fluorescent molecular rotor probe. Journal of Xenobiotics. 2019;**9**:8147-8149

[5] Verschoor A, ban Herwijnen R, Postuma C, Klesse K, Werner S. Assessment document of land-based inputs of microplastics in the marine environment. In: Environmental Impact of Human Activities Series. United Kingdom: OSPAR; 2017. https://www.ospar.org/documents?v=38018

[6] Lithner D, Larsson A, Dave G. Environmental and health hazard ranking and assessment of plastic polymers based on chemical composition. Science of the Total Environment. 2011;**409**:3309-3324

[7] Wang F, Bexiga MG, Anguissola S, Boya P, Simpson JC, Salvati A, et al. Time resolved study of cell death mechanisms induced by amine-modified polystyrene nanoparticles. Nanoscale. 2013;**5**:10868-10876

[8] Colvin VL, Kulinowski KM. Nanoparticles as catalysts for protein fibrillation. PNAS. 2009;**104**:8679-8680

[9] Gagné F, Auclair J, André C. Polystyrene nanoparticles induce anisotropic effects in subcellular fraction of the digestive system of freshwater mussels. Current Topics in Toxicology. 2019;**15**:43-49

[10] Sjollema SB, Redondo-Hasselerharm P, Leslie HA, Kraak MHS, Vethaak AD. Do plastic particles affect microalgal photosynthesis and growth? Aquatic Toxicology. 2016;**170**:259-261

[11] Canesi L, Ciacci C, Bergami E, Monopoli MP, Dawson KA, Papa S, et al. Evidence for immunomodulation and apoptotic processes induced by cationic polystyrene nanoparticles in the hemocytes of the marine bivalve Mytilus. Marine Environmental Research. 2015;**111**:34-40

[12] Cole M, Galloway TS. Ingestion of nanoplastics and microplastics by pacific oyster larvae. Environmental Science and Technology. 2015;**49**:14625-14632

[13] Cole M, Lindeque P, Fileman E, Halsband C, Galloway TS. The impact of polystyrene microplastics on feeding, function and fecundity in the marine copepod Calanus helgolandicus. Environmental Science and Technology. 2015;**49**:1130-1137

[14] Nasser F, Lynch I. Secreted protein eco-corona mediates uptake and impacts of polystyrene nanoparticles on Daphnia magna. Journal of Proteomics. 2016;**137**:45-51

[15] Canesi L, Ciacci C, Fabbri R, Balbi T, Salis A, Damonte G, et al. Interactions of cationic polystyrene nanoparticles with marine bivalve hemocytes in a physiological environment: Role of soluble hemolymph proteins. Environmental Research. 2016;**150**:73-81

[16] Parsai T, Kumar A. Understanding effect of solution chemistry on heteroaggregation of zinc oxide and copper oxide nanoparticles. Chemosphere. 2019;**235**:457-469

[17] Blaise C, Kusui T. Acute toxicity assessment of industrial effluents with a microplate-based *Hydra attenuata* assay. Environmental Toxicology and Water Quality. 1997;**12**:53-60

[18] Dańko MJ, Kozłowski J, Schaible R. Unraveling the non-senescence phenomenon in Hydra. Journal of Theoretical Biology. 2015;**382**:137-149

[19] Wilby OK. The Hydra regeneration assay. In: Proceedings of the Workshop Organised by Association Francaise de Teratologie. France; 1989. pp. 108-124

[20] Quinn B, Gagné F, Blaise C. *Hydra*, a model system for environmental studies. International Journal of Developmental Biology. 2012;**56**:613-625

[21] Murphy F, Quinn B. The effects of microplastic on freshwater *Hydra attenuata* feeding, morphology and reproduction. Environmental Pollution. 2018;**234**:487-494

[22] Bradford MM. A rapid and sensitive method for the quantitation of microgram quantities of protein utilizing the principle of protein-dye binding. Analytical Biochemistry. 1976;**72**:248-254

[23] Boryslawskyj M, Garrood AC, Pearson JT. Elevation of glutathione-s-transferase activity as a stress response to organochlorine compounds, in the freshwater mussel, *Sphaerium corneum*. Marine Environmental Research. 1988;**24**:101-104

[24] Gagné F. Oxidative stress (Chapter 6). In: Biochemical Ecotoxicology: Principle and Methods. New York: Elsevier Inc.; 2014. pp. 103-115

[25] Finney DJ. Statistical Methods in Biological Assay. 2nd ed. London, England: Griffin; 1964

[26] Zhang F, Wang Z, Wang S, Fang H, Wang D. Aquatic behavior and toxicity of polystyrene nanoplastic particles with different functional groups: Complex roles of pH, dissolved organic carbon and divalent cations. Chemosphere. 2019;**228**:195-203

[27] Liu Z, Yu P, Cai M, Wu D, Zhang M, Huang Y, et al. Polystyrene nanoplastic exposure induces immobilization, reproduction, and stress defense in the freshwater cladoceran Daphnia pulex. Chemosphere. 2019;**215**:74-81

[28] Mishra P, Vinayagam S, Duraisamy K, Patil SR, Godbole J, Mohan A, et al. Distinctive impact of polystyrene nano-spherules as an emergent pollutant toward the environment. Environmental Science and Pollution Research International. 2019;**26**:1537-1547

[29] Shao H, Han Z, Krasteva N, Wang D. Identification of signaling cascade in the insulin signaling pathway in response to nanopolystyrene particles. Nanotoxicology. 2019;**13**:174-188

[30] Jovanović B. Ingestion of microplastics by fish and its potential consequences from a physical perspective. Integrated Environmental Assessment and Management. 2017;**13**:510-515

Chapter 5

Application of Novel Technology in Aquaculture

Chung-Lun Lu, Shiu-Nan Chen and Shao-Wen Hung

Abstract

Aquaculture continues to be the fastest-growing food production field that has a lot of potential to meet the aqua-protein needs. The scientific and business communities are responding to the many challenges and opportunities inherent in the growing aquaculture field. Advancements in production and detection of harmful material systems and technologies are contributing to aquaculture industry's expansion and sustainability. All of these production system technologies are benefitting from expanding information and communication systems, which are enabling advances in every stage of production. In the future, the new farming operation of friendly environment will focus on the use of nonecology destructive substances, no antibiotics, and the natural probiotics or novel immunomodulatory substances to match the physiological regulation of cultured organisms and the management of aquaculture. The future scientific-based innovation will contribute toward meeting increasing food demands, while improving social, environmental, and financial sustainability of the global aquaculture industry.

Keywords: aquaculture, immunomodulatory, prevention management, probiotics, surface-enhanced Raman scattering

1. Introduction

As the catch of marine fisheries is decreasing year by year, the demand for aquaculture products in human food supply sources is increasing year by year. However, the rapid spread of aquatic diseases, reduced area of farmed land, or the decreased need of environmental conservation caused the deterioration of the global cultivation of aquatic animal environment. Now, the emphasis on ecological environmental protection and the risk of biological threats after the use of aquatic animal drugs are gradually proven and discovered.

In the future, the new farming operation of friendly environment will focus on the use of nonecology destructive substances, no antibiotics, and the natural probiotics or novel immunomodulatory substances to match the physiological regulation of cultured organisms [1] and the management of aquaculture [2–4], resulting in a good immune disease strategy for breeding organisms. Furthermore, rapid disease screening technology reduces the risk of disease spread that is prone to occur in manual operations.

2. Probiotics establish a balanced aquaculture environment management

2.1 Probiotics in aquaculture

The biggest challenge faced by the aquaculture industry is the malignant changes in the farming environment, whether it is changed by man-made or natural conditions. Aquaculture farming environmental degradation often leads to the accumulation of toxic substances or the dominance of pathogenic microorganisms. Since the industrial operator has only paid attention to the improvement of technology and production yield for a long time, and neglected how to maintain the ecological balance of the culture pond and the surrounding environment, when a disease breaks out, the first strategy is to apply animal drugs, especially antibiotics; undeniably, drug treatment for a disease is the most direct and immediate practice [5], but its damage to the environment is difficult to estimate. In the aquaculture industry, the negative impact of the application of drugs on the environment has received attention increasingly. In addition, drug residues cause food safety concerns and may lead pathogenic microorganisms to become drug resistant [5], which has an irreversible effect on nutrients in the culture environment [6].

So, the main purpose of adding probiotics or microbial products in the aquaculture pond is to decompose organic substances and at the same time to inhibit the growth of pathogenic microorganisms, reduce the risk of disease occurrence [7], convert toxic substances, reduce the accumulation of organic matter, stabilize the water color, and balance ecosystem in water. In the early days, the definition of probiotics was limited to microorganisms that were beneficial to the health of the organism. With the continuous revision, the current broad definition of probiotics not only increases the health of the organism but also improves the survival rate and the quality of the water, improving the immune mechanism of animals and even stabilizing the balance of the bacteria in the body [8], which are all functions of probiotics, and these benefits are very important for the aquaculture process, especially when the fry of aquatic animal are incubated, using probiotics as microbial additives to improve the final aquaculture production yield has important benefits [9].

Microbial additives usually do not have only a single strain. A single strain will have some disadvantages in the treatment of the aquaculture environment. A single strain usually only has the ability to handle certain substances, and the organic components in the aquaculture water are extremely complicated. Although the single strain treatment will change the figures or water quality numbers in the research, the improvement of the overall water quality is not great, so the addition of compound strains has become one of the important concepts of microbial additives. It is one of the important goals of microbial additives to treat different wastes in aquaculture environment with different strains to achieve the goal of purifying and stabilizing the whole culture system. The classification of probiotics in culture is distinguished by the currently widely used probiotics, which can be distinguished by their efficacy and timing of use: as follows: (i) probiotics for fry incubation (as show in **Table 1**); (ii) probiotics for cultured adult or subadult organisms (as show in **Table 2**); (iii) antipathogenic probiotics screened in the environment (as show in **Table 3**); and (iv) probiotics that increase the immune mechanism for aquatic animals.

The immune mechanisms of many aquatic animals are not as complete as those of higher mammals, and most aquatic products rely heavily on nonspecific immune mechanisms due to the lack of an active mechanism for antibody immunity. In the case of shrimps, in general, nonspecific immune mechanisms mainly include phagocytosis in the blood cells, nodule formation, encapsulation, and the interaction of other substances in the plasma, which include the cytotoxicity of cytotoxins, complement activated by the action of lectin, and the role of proPO system and

Species of probiotics	Aquaculture animal
Thalassobacter utilis PM4	Shrimp (*Penaeus monodon*)
Trosomonas sp.	Shrimp (*Penaeus monodon* and *Penaeus penicillatus*)
Nitrobacter sp.	Shrimp (*Penaeus monodon* and *Penaeus penicillatus*)
Bacillus sp.	Shrimp (*Penaeus monodon* and *Penaeus penicillatus*)
Thalassobacter utilis PM4	Crab (*Portunus trituberculatus*)
Bacillus toyoi	Turbot-via rotifers (*Brachionus plicatilis*)
Bacillus sp. spores	Turbot-via rotifers (*Brachionus plicatilis*)
Lactic bacteria	Turbot-via rotifers (*Brachionus plicatilis*)
Lactobacillus plantarum	Turbot-via rotifers (*Brachionus plicatilis*)
Lactobacillus helveticus	Turbot-via rotifers (*Brachionus plicatilis*)
Lactobacillus balgaricus	Turbot-via rotifers (*Brachionus plicatilis*)
Streptococcus lactis	Turbot-via Artemia
Alteromonas sp.	Oyster *(Crassostrea gigas)*
Roseobacter sp.(BS107)	Scallop (*Pecten maximus*)

Table 1.
Probiotics for aquatic animal fry incubation.

antibacterial protein/peptides [10, 11]. The biggest disadvantages of these effects are in two parts: one is the lack of memory, and the biggest disadvantage of the mechanism of nonspecific immunity is this. Due to the lack of memorable specificity, the phenomenon of repeated infection becomes a big burden in the aquaculture breeding industry. The second disadvantage is that some immune functions usually do not start real time and they need some special activating substances, such as the

Species of probiotics	Aquaculture animal	Effect
Lyoohilized *Carnobacterium divergens*	Atlantic salmon	Decrease *Vibrio anguillarm* strain infection
Lyoohilized *Carnobacterium divergens*	Atlantic salmon	Decrease *Aeromonas salmonicida* infection
Carnobacterium strain K1	Flounder	Inhibition the proliferation of *V. anguillarm* and *A. salmonicida*
Carnobacterium	Flounder	Inhibition the proliferation of V. anguillarm

Table 2.
Probiotics for cultured adult or sub-adult aquatic organisms.

Species of probiotics	Aquaculture animal	Effect
Nitrosomonas sp.	Shrimp fry	Increase survival rate
Nitrobacter sp.	Shrimp fry	Increase survival rate
Bacillus sp	Fry and adult fish	Increase survival rate
Spray-dried *Tetraselmis suecica*	Atlantic salmon adult fish	Increase the survival rate after poly-pathogenic infection

Table 3.
Anti-pathogenic probiotics screened in the environment.

participation of immune-stimulator to initiate the original immune state, so the reaction time is much delayed. This is also one of the reasons for the sudden death of the aquaculture animals.

Recently, many studies have found that the addition of probiotics can improve the immunity of cultured organisms. Such a phenomenon is very helpful for aquaculture organisms. Especially for the cultured organisms of shrimps and crabs, because they mainly rely on nonspecific immune mechanisms, how to activate their nonspecific immune system and improve their efficacy has become an important issue in aquaculture management. According to the report, using the source of immune activation, substances such as β-glucan can significantly stimulate the nonspecific immune mechanism of many water-generating substances, and these related polysaccharides are separated from the secondary metabolites of microorganisms. Such results have led to many ideas for making related microorganisms into biological additives, and many studies have confirmed that active yeasts and polysaccharides produced by beneficial bacilli do have positive effects on improving the survival rate and yield of cultured organisms. Feed was added with probiotics or directly added with polysaccharides, and positive benefits in improving the disease resistance and survival rate of cultured organisms have been found [12, 13].

2.2 Effects of probiotics on aquaculture environmental indicators

At present, probiotics are being used in the treatment of industrial wastewater, domestic sewage, animal waste, soil improvement, etc., and their functions in the aquaculture environment are as follows: (i) Inhibiting the occurrence of diseases: beneficial microorganisms can inhibit the growth of pathogenic bacteria by secreting inhibitory substances, competing nutrients, and competing for attachment sites. (ii) Maintaining good water quality: adding beneficial microorganisms can promote the decomposition of organic matter or the conversion of toxic substances, reduce the accumulation of organic matter in the culture pond, and convert toxic substances into nutrient compounds with low toxicity or even reusability, such as ammonia. It is converted to nitrous acid by nitrification and then converted to nitric acid, which can be used for the growth and utilization of plant plankton, which has the effect of stabilizing water quality and balancing the ecosystem. (iii) Improving health and growth: probiotics can improve the bacterial phase in the digestive tract of cultured organisms, promote the digestion and absorption of cultured organisms, help the health and growth of cultured organisms, and increase the survival rate to increase production yield. (iv) Increasing natural bait: the use of photosynthetic bacteria to culture animal plankton can increase the survival rate of brine shrimp, and the added bacteria can be used as a food for protozoa, and the bacteria themselves are a good source of bait to reduce the use of artificial feed.

Probiotics are used in aquaculture environments in the following ways: (i) combined with artificial feed, (ii) put in culture pond water, (iii) immersion, (iv) via live baits, and (v) via medium release. The usage timing can be divided into preculture and postculture. The former is used as the whole pool rearranges with the hope of utilizing the benefit bacteria phase, while the latter is used as the maintenance of the stable culture environment. The additives in feed can be divided into mixed and mixed for fermentation. It is added after fermentation to increase the digestion and absorption rate of the organisms, thereby reducing the excretion and pollution of the water. The addition of biological wastewater treatment system can be regarded as the microbial planting in the water system, which can ensure the stability of the water quality of the system and increase the effect of treatment. The easiest way to use probiotics to treat water quality is to put the microbial products directly into the culture pond. However, this kind of externally introduced probiotics must be made into a dominant species to function, so it must be continued every other day. To achieve the desired results, the probiotics mass and quantity of the products used should be especially confirmed. In addition, the effect of probiotics is quite susceptible to environmental factors such as salinity, temperature, and pH value. Therefore, it is best to activate the probiotics first and

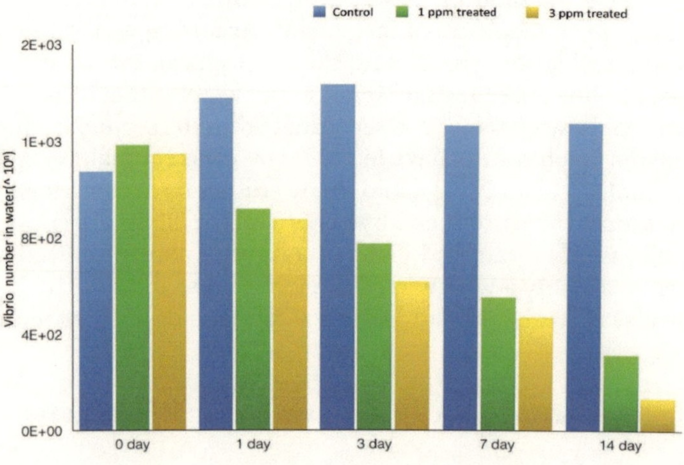

Figure 1.
Effectiveness of photosynthetic bacteria applied in white shrimp culture. Change in the total amount of Vibrio *in the culture pond within 14 days of continuous use of photosynthetic bacteria in the white shrimp culture process. The results showed that in the experimental group with more than 1 ppm of photosynthetic bacteria added continuously, the total amount of* Vibrio *was significantly reduced, which proved that the use of photosynthetic bacteria can effectively reduce the number of* Vibrio *in water.*

Water quality index	Before use	After use
BOD	82 ppm	16 ppm
COD	59 ppm	7 ppm
NH4-N	8 ppm	1.5 ppm
H2S	1.5 ppm	0.03 ppm

Table 4.
Probits stabilizes the water quality, but also reduces the accumulation of residual.

adjust the temperature of the mixture to be similar to the temperature of the pool water, and because the probiotics themselves are live bacteria, they should not be mixed with antibacterial agents, sulfonamides, antibiotics, and disinfectant water or in high temperature process.

An example of our use in white shrimp (*Litopenaeus vannamei*) culture farms, we use a variety of special strains such as Photosynthetic bacteria *Rhodobacter capsulatus* and *Micrococcus luteus*. Photosynthetic bacteria can utilize the hydrogen sulfide at the bottom of the pool as a nutrient to remove harmful substances and absorb heavy metals and inhibit the growth of Vibrio (**Figure 1**), while *M. luteus* can rapidly decompose proteins and increase the speed of nitrogen circulation in water to effectively reduce the occurrence and damage of sediments at the bottom of the pool. The interactive use of different strains not only effectively stabilizes the water quality but also reduces the accumulation of residual bait and organic matter in the shrimp culture process (**Table 4**) and greatly inhibits the proliferation of pathogenic bacteria such as Vibrio in the water. Such technology has been widely used in aquaculture industry in Taiwan and in Surabaya, Indonesia, and Thailand. Photosynthetic bacteria have become one of the most beneficial microbial strains in today's aquaculture industry (**Figure 2**).

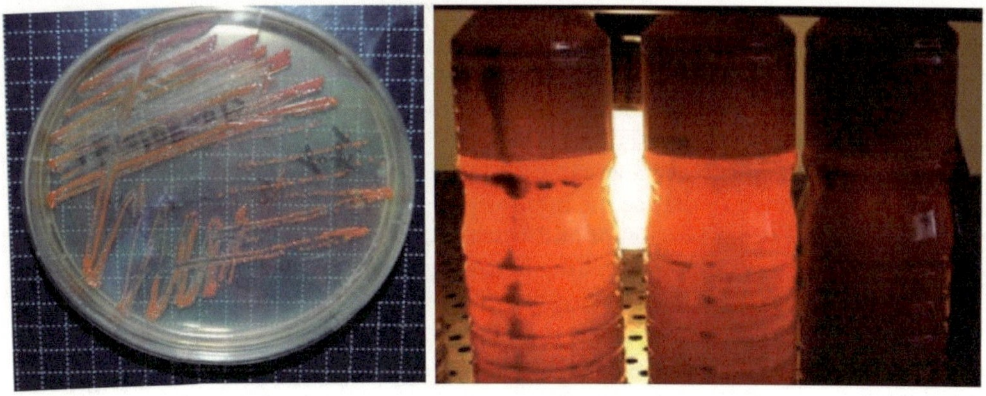

Figure 2.
Photosynthetic bacteria culture and mass production. Purification of photosynthetic bacteria on agar plate (left). Commercial mass production of photosynthetic bacteria (right).

3. Strategies for cultured aquatic animals' immunization and immune regulation

3.1 Aquatic animals' immunity and immune regulation

Animal congenital immune cells or other nonimmune cells, such as epithelial cells that secrete cytokines, mononuclear cells, macrophages, dendritic cells, and natural killer cells, have many nonspecific pattern recognition receptors, targeting identification and binding of antigen-related molecular patterns on invading pathogens. Complement receptor type 3 (CR3 receptor) on these cells is a receptor for the recognition of β-glucan [14], and when β-glucan binds to CR3 receptor, it triggers a series of signaling cascades to activate transcription factors, which translates cellular transcription into cytokines that trigger inflammatory responses, and incrementally regulates the expression of the major histocompatibility complex (MHC) of the antigen. In turn, other immune cells are activated to achieve immunomodulatory functions [15].

The polysaccharide extracted from the yeast (*Saccharomyces cerevisiae*) was orally administered for 4 weeks in dogs. The expression of total IgA, IgM, and IgG in serum and lacrimal gland was detected, and the expression of IgA in serum was found to be significantly decreased and IgM increased significantly, while IgG did not change significantly; the amount of IgA in the lacrimal gland was also significantly reduced. In the second part, the Bordetella vaccine was injected subcutaneously, and the specific IgA, IgM, and IgG showed the same trend as in the first part. Oral administration of β-glucan to Miguel dog changed the expression of IgA and IgM in serum, and there was no difference in the expression of IgG, indicating that the intestinal mucosal immune response was stimulated through this way. Glucan enters the gut-associated lymphoid tissue from M cells and binds to dentin 1 and TLR-2 on macrophages and dendritic cells, secreting IL-12 and TNF-α and other cytokines, which alter the cytokine microenvironment to stimulate the homologous transformation of B cells, affecting the secretion of immunoglobulins of different isoforms [16]. Dendritic cells are one of the important antigen-presenting cells in the immune system. Lu et al. used a polysaccharide extracted from *Antrodia camphorate* to carry out cell experiments and cocultured it with immature dendritic cells and T cells isolated from healthy human blood to investigate the maturation of dendritic cells and T-cell proliferation. The result showed polysaccharides promote maturation of dendritic cells to stimulate T-cell proliferation and IFN-γ expression [17]. Coculture of polysaccharides with macrophages can promote macrophage secretion of immune-related factors and cytokine gene expression, such as nitric oxide (NO), tumor necrosis factor-α (TNF-α), IL-1β, IL-6, etc., and promote macrophage activity [18].

According to the description of the fish immune system, we believe that fish do have a memory-specific immune function, so the administration of the vaccine should be quite feasible, but it is considered that the vaccine must be applied to the suitable stage and subject that can cause the immune reaction, and also regular additional injections to enhance the memory effect for the immune response is necessary. However, the application of additional injections is a slight inconvenience for the user who works on the breeding aquaculture farms. Therefore, the immunomodulatory substances can be used complementarily to enhance the nonspecific immune function and to resist pathogens quickly in the farmed fish. Currently, we are investing in the research and development of immunomodulatory substances, in order to solve the dilemma of the decline of biological and physiological ability of breeding. In recent years, research on immunomodulatory technology or the development of immunomodulatory products has gradually become a prominent feature in aquaculture disease control and health management strategies. Many studies have found that once the nonspecific immunity and specific immune function of the cultured organisms are effectively improved, the survival rate, growth rate, and disease resistance of the cultured organisms are significantly improved. It is a pity that many immunomodulatory substances incur excessively high production costs or inherent unitability that it was difficult to obtain similar benefits as the research results in farm operations. In order to solve such problems, our research team spent more than 10 years of painstaking efforts to find the best solution in the macromolecular polysaccharides extracted from mushroom. We overcame the bottleneck of poor stability and high production cost of polysaccharide products in the past and made the immunomodulatory additives to be more easily applied in aquaculture.

3.2 Immunomodulatory additives in aquaculture

Mushroom glucan (MBG) is a kind of natural polymeric substance (polysaccharide), which is an immunomodulatory substance, and therefore does not cause an uncontrolled increase in immunity in fish, but certain drugs, such as levamisole, may cause undesirable side effects.

In a case study on eel cultured in fresh water, we have found that MBG can properly regulate the cellular immunity; the head, kidney, and spleen are the most important hematopoietic tissues of eel, and when the fish is infected with bacteria or viruses, blood cells from hematopoietic tissues will suffer damage, its number will also be significantly reduced, and the results show that the polysaccharide applied will significantly activate the hematopoietic system of the eel, effectively increasing the number of leukocytes. After 48 hours of using the MBG, the phagocytic cells with phagocytic ability in the body can be increased by about 4–6 times; experimental observations show that the phagocytosis ability is also significantly increased by the polysaccharide-activated phagocytic cells. Studies have confirmed that after the use of polysaccharides, they can significantly enhance the cellular immune function of eels, so that the fish body can resist pathogen infection, which is effective for improving health and improving disease resistance.

The immune stimulatory effects of immunostimulants like glucan, chitosan, and other polysaccharides have been widely studied in fish and crustaceans and were reviewed by Sakai [19]. When fish received immunomodulatory substance, macrophage undergoes phagocytosis, and the rate of intracellular oxygen consumption also increases at the same time, resulting in a so-called respiratory burst phenomenon, and the reaction of oxygen and NADPH into a reactive oxygen species with bacteriostatic ability via NADPH oxidase such as superoxide anion (O_2^-) and then will be converted to hydrogen peroxide (H_2O_2) by superoxide dismutase (SOD). The produced hydrogen peroxide can be transferred to hypochlorous acid (HOCl), etc., by myeloperoxidase. The cells can use these reactive oxygen intermediates (ROIs) to avoid the invasion of pathogens [20].

In the study of cobia (*Rachycentron canadum*) fish, it was found that after treatment with polysaccharides, the leukocyte phagocytosis rate, phagocytic index, superoxide ion yield, and lysozyme-like immune-related index were significantly improved in the 2–4 days after treatment. The activity of superoxide dismutase was highest in 4–8 days, or even up to 30 days, similar to that in the untreated group in the physiological index, thus demonstrating the efficient use of polysaccharide in the culture of cobia. It can effectively stimulate macrophages in a short period of time, improve the immunity of fish, and provide a continuous supply of immune-stimulating activity through the addition of feed, which is very helpful for the defense against disease and disease outbreak stage. And long-term treatment has no adverse effects on the fish. For the spot grouper, which is currently the main cultured seawater economic fish in Taiwan, the research has similar effects, and polysaccharides can be used for a long term and be increased in the treatment with 0.8 g/kg and 1.6 g/kg feed. The effect of phagocytes and other related nonspecific immune responses also increases significantly.

In invertebrates' culture industry, the abalone culture process often suffered problems in low rate of attachment and poor resistance to foreign pathogens. In abalone immune system, pattern recognition proteins (PRPs) play an important role in innate immunity by recognizing common epitopes on the surface of invading microorganism. In a study of pathogenic challenge experiment of abalone, the test groups were injected with Vibrio, lipopolysaccharides (LPS), and beta-1,3-glucan, respectively, and then compared to controls. Abalone PRP increase can recognize different pathogen-associated molecular patterns (PAMP) and may activate different genes involved in the defense against these pathogens. It acts as an acute inducible protein that could play an important role in the abalone immune defense mechanism [21].

The shrimp immune system has to be activated by pattern recognition proteins such as lipopolysaccharide, β-glucan, and peptidoglycan-binding proteins [22, 23]. Once these proteins are bound to their particular targets, they activate hemocytes to release their contents and trigger different biochemical mechanisms [24]. Immunostimulants increase the immune responses in several shrimp species by promoting phagocytosis, bactericidal activity, proPO activity, and respiratory bursts and enhancing resistance against pathogens [25, 26].

The use of immersion, injection, and feeding with polysaccharides to improve the nonspecific immune response for aquaculture organisms has become popular. This polysaccharide efficacy was found in the administration with freshwater carp, catfish, trout, eel, tilapia, or seawater salmon, flounder, grouper, cobia, etc. Finally, although the effects of polysaccharide function on aquatic animal-specific immunity are not directly stimulated and promoted, in other animal experiments, it has been found that the use of polysaccharides as an adjuvant for animal vaccines can increase the vaccine potency, stimulating the production of antibodies. The vaccine of fish has also been widely studied in recent years. With the aid of polysaccharides, it is believed that more research and certification are needed. Therefore, it is hoped that this will be another culture technology in the near future prospect.

4. Novel disease screening technology

The Raman spectrometer (RS) is an instrument that measures the Raman scattering light spectrum. Raman scattered light was discovered by Indian scientists in 1928. Unlike the general laser scattered light, the wavelength of Raman scattered light is slightly different from the wavelength of the original incident light. The reason for the difference is that the collision occurs after the molecules are hit by photons. Increasing or reducing photon energy causes the effects of these changes due to molecular bonding and structure [27]. Therefore, this technology is widely used in the detection of high molecular polymers, nanomaterials, electrochemistry, semiconductors, thin films, mineralogy, carbides, etc.

Surface-enhanced Raman scattering (SERS) has been proven to be highly sensitive for trace chemical detection. The power of SERS detection is the label-free detection capabilities, due to the highly specific Raman scattering fingerprint spectrum (RSFS) and combined with the high sensitivity. SERS has been used to detect a wide range of targets including bio- and nonbiomaterials (DNA, proteins, pesticides, and metabolites). RSFS is a narrow band signal with highly specific characteristic to imply that it can be used to detect multiple analytes simultaneously. Therefore, SERS may finally realize its potential as a highly sensitive and specific analytical technique for trace chemical and biomolecule detection [28].

In recent years, due to the advancement of CCD and laser, scientists have gradually pulled RS to use in the detection of bioassays and medical drugs [27, 29]. At present, SERS is used to enhance the signals needed for specific analysis. It has been widely used in life science research. Common applications include pharmaceuticals, clinical trials, cellular research, immunology, proteomics research, genetics, genetic engineering, plastic surgery, biomedical materials, environmental engineering, and biosafety. In the field of life science research, SERS is able to identify samples with single cell-level accuracy. In the future, instruments combining Raman spectroscopy and microscopic imaging technology will bring great benefits to the life science community.

With the intensive and increasing breeding scale of pigs in various countries, the problem of pig diseases is becoming more and more complicated, and mixed infections or secondary infections are also increasing. Therefore, the diagnosis and prevention of diseases have increasingly shown their importance. In addition, due to the use of antibiotics and vaccines, the symptoms of bacterial (porcine mycoplasma pneumonia and Streptococcus suis infection) or viral (porcine reproductive and respiratory syndrome and porcine type 2 circovirus infection) infections are atypically present. They also often cause serious economic losses for the livestock industry. If they are diagnosed by relying only on clinical symptoms, it will be difficult to confirm the diagnosis. Additionally, the need to strengthen laboratory diagnosis is currently the primary choice. Therefore, it is necessary to develop a nonintrusive real-time image monitoring system. In our laboratory, we tried to characterize and evaluate a SERS-based diagnostic system for the detection and identification of bacteria in the gnotobiotic mice for rapid detection of two common kinds of bacteria in rodent (*Staphylococcus aureus* and *Pseudomonas aeruginosa*)

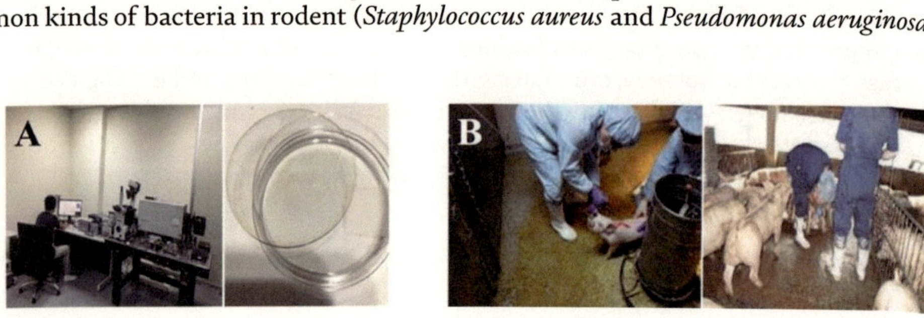

Figure 3.
SERS-based detection system and pig sample collections. (A) the SERS-based detection system. (B) Sample collections from the traditional pig farms.

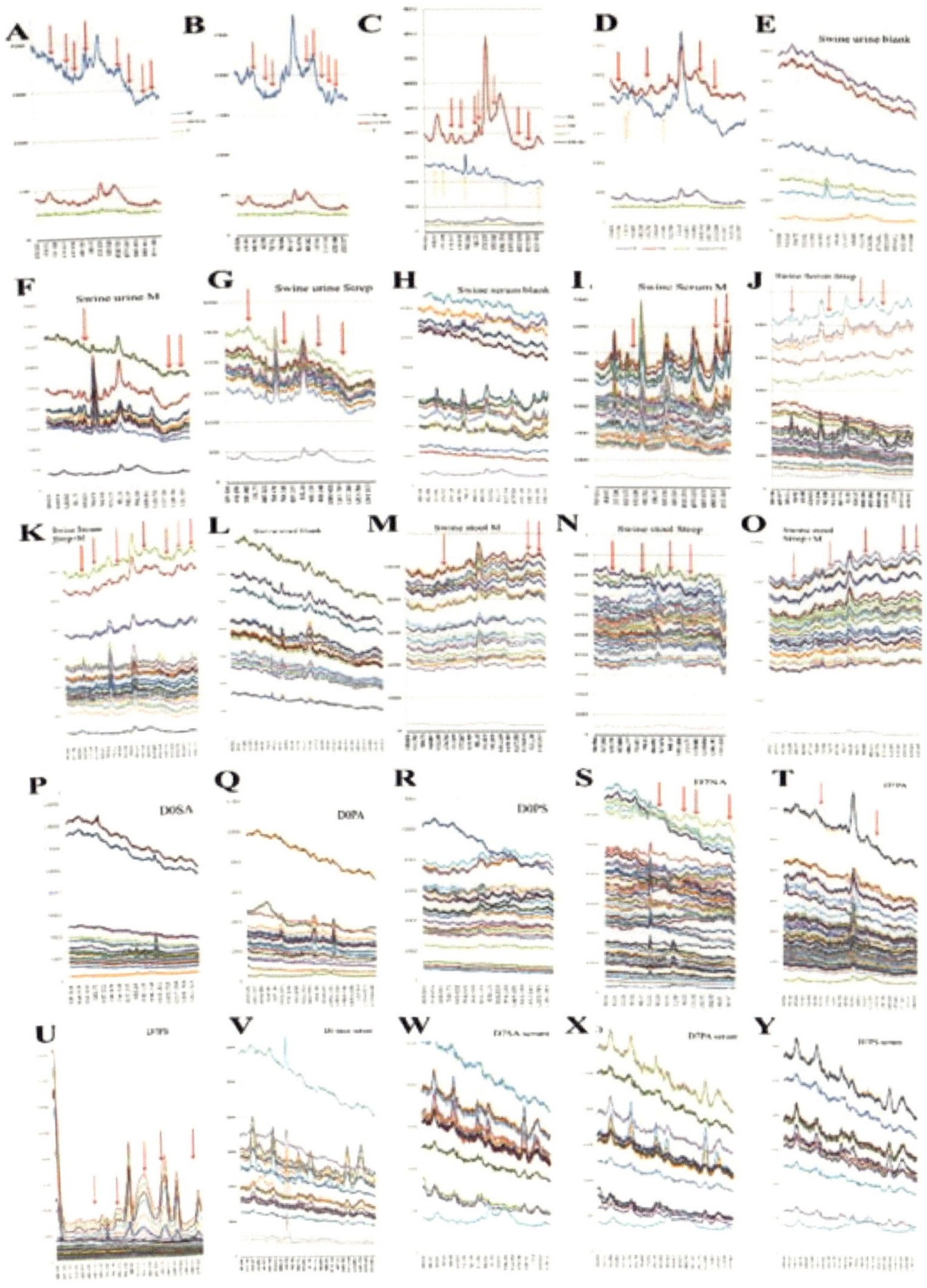

Figure 4.
SERS spectra obtained with (A) Mycoplasma hyopneumoniae, *(B)* Streptococcus suis, *(C) Staphylococcus aureus, and (D) Pseudomonas aeruginosa. Bacteria recovered from pooled swine urine: (E) swine urine, (F) M. hyopneumoniae, and (G)* S. suis. *Bacteria recovered from pooled swine serum: (H) swine serum, (I) M. hyopneumoniae, (J)* S. suis, *(K) M. hyopneumoniae, and* S. suis. *Bacteria recovered from pooled swine feces: (L) swine feces, (M) M. hyopneumoniae, (N)* S. suis, *(O) M. hyopneumoniae, and* S. suis. *Colonization of gnotobiotic mice with the known bacterial strain: 0 day (feces specimens) (P) S. aureus, (Q) P. aeruginosa, (R) S. aureus, and P. aeruginosa. 7 days after colonization (feces specimens) (S) S. aureus, (T) P. aeruginosa, (U) S. aureus, and P. aeruginosa. 0 day (serum specimens) (V) without colonization. 7 days after colonization (serum specimens) (W) S. aureus, (X)* P. aeruginosa, *(Y) S. aureus, and P. aeruginosa. Arrows indicate the highly specific peaks of Raman scattering fingerprint spectrum for pathogens.*

and two kinds of pathogenic bacteria in swine (Mycoplasma hyopneumoniae and Streptococcus suis). SERS spectra of bacteria recovered from sera, feces, and urine were compared to pure culture bacteria. These results indicated that successful detection, identification, and classification of these bacteria from mice specimens (sera and feces) and swine specimens (sera, feces, and urine) using a SERS-based system were demonstrated. SERS is shown to offer reproducible molecular spectroscopic signatures for analytical applications. The approach may be a new and powerful tool for real-time surveillance of animal bacterial disease pathogens in clinics (**Figures 3** and **4**) [28].

This shift away from traditional aquaculture farming to large-scale intensive methods has resulted in the increasing cases of complex, mixed infections, or secondary infections. The early diagnostic and disease prevention becomes more important for aquaculture farm management. However, antibiotic treatment caused atypical diseases symptoms might be interfered clinical diagnosis. In order to perform accurate pathological diagnosis, we need to combine the strengthening laboratory diagnosis and traditional diagnostics. SERS has recently been shown to be a potentially powerful whole-organism fingerprinting technique and is used for the rapid identification of bacteria. Biosensors based on SERS hold great promise as a platform for sensitive and rapid detection of bacterial pathogens and decrease time of diagnosis [30].

In spite of these capabilities, SERS has been limited to research labs due to stationary equipment and high cost substrates. The instruments combining Raman spectroscopy and microscopic imaging technology will continuously be developed. At present, a highly sensitive and rapid method of SERS combining with electrochemical preconcentration in detecting malachite green in aquaculture water has been established [31]. However, application of SERS in the aquaculture pathogen detections is lacking now. In the future, the application of SERS in the rapid detection for aquaculture pathogens or veterinary medical drugs in the aquaculture farms is very important and needed.

5. Conclusions

Aquaculture is the fastest-growing food production field. The scientific and business communities are responding to the many challenges and opportunities. The new farming operation of friendly environment will focus on the use of nonecology destructive substances, no antibiotics, and the natural probiotics or novel immunomodulatory substances to match the physiological regulation of cultured organisms and the management of aquaculture. Finally, R&D of the novel pathogen detection technology in the aquaculture is also very important and needed.

Acknowledgements

The authors thank the Ministry of Science and Technology and the Council of Agriculture, Taiwan, for support.

Conflict of interest

The authors declare no conflict of interest.

DOI: http://dx.doi.org/10.5772/intechopen.90142

Author details

Chung-Lun Lu[1], Shiu-Nan Chen[2] and Shao-Wen Hung[3]*

1 Aquatic Technology Laboratories, Agricultural Technology Research Institute, Hsinchu, Taiwan

2 Institute of Fisheries Science, National Taiwan University, Taipei, Taiwan

3 Animal Technology Laboratories, Agricultural Technology Research Institute, Hsinchu, Taiwan

*Address all correspondence to: 1032169@mail.atri.org.tw

References

[1] Mohapatra S, Chakraborty T, Prusty AK, PaniPrasad K, Mohanta KN. Beneficial effects of dietary probiotics mixture on hemato-immunology and cell apoptosis of Labeo rohita fingerlings reared at higher water temperatures. PLoS One. 2014;**9**:e100929

[2] Costa-Pierce BA. Sustainable ecological aquaculture systems: The need for a new social contract for aquaculture development. Marine Technology Society Journal. 2010;**44**:88-112

[3] Neori A, Chopin T, Troell M, Buschmann AH, Kraemer GP, Halling C, et al. Integrated aquaculture: Rationale, evolution and state of the art emphasizing seaweed biofiltration in modem mariculture. Aquaculture. 2004;**231**:361-391

[4] Lamari F, Mahdhi A, Chakroun I, Esteban MA, Mazurais D, Amina B, et al. Interactions between candidate probiotics and the immune and antioxidative responses of European sea bass (Dicentrarchus labrax) larvae. Journal of Fish Diseases. 2016;**39**:1421-1432

[5] Ma X, Zhu F, Jin Q. Antibiotics and chemical disease-control agents reduce innate disease resistance in crayfish. Fish & Shellfish Immunology. 2019;**86**:169-178

[6] Schmidt AS, Bruun MS, Dalsgaard I, Pedersen K, Larsen JL. Occurrence of antimicrobial resistance in fish-pathogenic and environmental bacteria associated with four danish rainbow trout farms. Applied and Environmental Microbiology. 2000;**66**:4908-4915

[7] Rengpipat S, Tunyanun A, Fast AW, Piyatiratitivorakul S, Menasveta P. Enhanced growth and resistance to vibrio challenge in pond-reared black tiger shrimp Penaeus monodon fed a bacillus probiotic. Diseases of Aquatic Organisms. 2003;**55**:169-173

[8] Tapia-Paniagua ST, Chabrillon M, Diaz-Rosales P, de la Banda IG, Lobo C, Balebona MC, et al. Intestinal microbiota diversity of the flat fish Solea senegalensis (Kaup, 1858) following probiotic administration. Microbial Ecology. 2010;**60**:310-319

[9] Vine NG, Leukes WD, Kaiser H. Probiotics in marine larviculture. FEMS Microbiology Reviews. 2006;**30**:404-427

[10] Johansson MW, Lind MI, Holmblad T, Thornqvist PO, Soderhall K. Peroxinectin, a novel cell adhesion protein from crayfish blood. Biochemical and Biophysical Research Communications. 1995;**216**:1079-1087

[11] Antony SP, Singh IS, Sudheer NS, Vrinda S, Priyaja P, Philip R. Molecular characterization of a crustin-like antimicrobial peptide in the giant tiger shrimp, Penaeus monodon, and its expression profile in response to various immunostimulants and challenge with WSSV. Immunobiology. 2011;**216**:184-194

[12] Sivakamavalli J, Vaseeharan B. Purification, characterization and functional analysis of a novel beta-1, 3-glucan binding protein from green tiger shrimp Penaeus semisulcatus. Fish & Shellfish Immunology. 2013;**35**:689-696

[13] Zokaeifar H, Babaei N, Saad CR, Kamarudin MS, Sijam K, Balcazar JL. Administration of *Bacillus subtilis* strains in the rearing water enhances the water quality, growth performance, immune response, and resistance against Vibrio harveyi infection in juvenile white shrimp, Litopenaeus vannamei. Fish & Shellfish Immunology. 2014;**36**:68-74

[14] Vetvicka V, Yvin JC. Effects of marine beta-1,3 glucan on immune reactions. International Immunopharmacology. 2004;**4**:721-730

[15] Pashine A, Valiante NM, Ulmer JB. Targeting the innate immune response with improved vaccine adjuvants. Nature Medicine. 2005;**11**:S63-S68

[16] Stuyven E, Verdonck F, Van Hoek I, Daminet S, Duchateau L, Remon JP, et al. Oral administration of beta-1,3/1,6-glucan to dogs temporally changes total and antigen-specific IgA and IgM. Clinical and Vaccine Immunology: CVI. 2010;**17**:281-285

[17] Lu MC, Du YC, Chuu JJ, Hwang SL, Hsieh PC, Hung CS, et al. Active extracts of wild fruiting bodies of Antrodia camphorata (EEAC) induce leukemia HL 60 cells apoptosis partially through histone hypoacetylation and synergistically promote anticancer effect of trichostatin a. Archives of Toxicology. 2009;**83**:121-129

[18] Ljungman AG, Leanderson P, Tagesson C. (1-->3)-beta-d-Glucan stimulates nitric oxide generation and cytokine mRNA expression in macrophages. Environmental Toxicology and Pharmacology. 1998;**5**:273-281

[19] Sakai M, Kono T, Savan R. Identification of expressed genes in carp (Cyprinus carpio) head kidney cells after in vitro treatment with immunostimulants. Developmental Biology (Basel). 2005;**121**:45-51

[20] Roszell LE, Anderson RS. In vitro immuno modulation by pentachlorophenol in phagocytes from an estuarine teleost, Fundulus heteroclitus, as measured by chemiluminescence activity. Archives of Environmental Contamination and Toxicology. 1993;**25**:492-496

[21] Nikapitiya C, De Zoysa M, Lee J. Molecular characterization and gene expression analysis of a pattern recognition protein from disk abalone, Haliotis discus discus. Fish & Shellfish Immunology. 2008;**25**:638-647

[22] Cheng W, Liu CH, Tsai CH, Chen JC. Molecular cloning and characterisation of a pattern recognition molecule, lipopolysaccharide- and beta-1,3-glucan binding protein (LGBP) from the white shrimp Litopenaeus vannamei. Fish & Shellfish Immunology. 2005;**18**:297-310

[23] Roux MM, Pain A, Klimpel KR, Dhar AK. The lipopolysaccharide and beta-1,3-glucan binding protein gene is upregulated in white spot virus-infected shrimp (Penaeus stylirostris). Journal of Virology. 2002;**76**:7140-7149

[24] Wang XW, Wang JX. Pattern recognition receptors acting in innate immune system of shrimp against pathogen infections. Fish & Shellfish Immunology. 2013;**34**:981-989

[25] Campa-Cordova AI, Hernandez-Saavedra NY, De Philippis R, Ascencio F. Generation of superoxide anion and SOD activity in haemocytes and muscle of American white shrimp (Litopenaeus vannamei) as a response to beta-glucan and sulphated polysaccharide. Fish & Shellfish Immunology. 2002;**12**:353-366

[26] Chang CF, Chen HY, Su MS, Liao IC. Immunomodulation by dietary beta-1, 3-glucan in the brooders of the black tiger shrimp Penaeus monodon. Fish & Shellfish Immunology. 2000;**10**:505-514

[27] Luo SC, Sivashanmugan K, Liao JD, Yao CK, Peng HC. Nanofabricated SERS-active substrates for single-molecule to virus detection in vitro: A review. Biosensors & Bioelectronics. 2014;**61**:232-240

[28] Hung RC, Chen CC, Lin JW, Chuang HL, Chiu CC, Huang PM, et al. Development of rapid detection of animal pathogens by the surface-enhanced Raman scattering system. Journal of Agriculture and Forestry, National Chiayi. 2019;**16**:61-84

[29] Mayne ST, Cartmel B, Scarmo S, Jahns L, Ermakov IV, Gellermann W. Resonance Raman spectroscopic evaluation of skin carotenoids as a biomarker of carotenoid status for human studies. Archives of Biochemistry and Biophysics. 2013;**539**:163-170

[30] Hung YW, Lin YH, Chen MH, Wang WS, Chiu CF, Chiu CC, et al. Pharmacokinetic study of florfenicol in bester sturgeon, a cultured hybrid of *Huso huso* × *Acipenser ruthenus* by high performance liquid chromatography equipped with UV detector. Aquaculture. 2018;**495**:558-567

[31] Xu KX, Guo MH, Huang YP, Li XD, Sun JJ. Rapid and sensitive detection of malachite green in aquaculture water by electrochemical preconcentration and surface-enhanced Raman scattering. Talanta. 2018;**180**:383-388

Chapter 6

Development of Ex Vivo Model to Study the Effect of Rock Snot, *Didymosphenia geminata* (Lyngb.), on Cells and Gametes of Freshwater Fish

Jorge Parodi, Viviana Chavez and Matias Peredo-Parada

Abstract

Rock snot, a species of diatom that produces nuisance growths in international freshwater rivers and streams, in Chile, Canada, New Zealand and other country, with consistently cold-water temperatures and low-nutrient levels, is produced by *Didymosphenia geminata* (Lyngb.) M. Schmidt (*D. geminata*) and is a complex problem in rivers. Its biology problem and its effects on ecosystems are still being investigated, yet no research has focused on the effects of *D. geminata* at the cellular level. We proposed to review and described an artificial river system to preserve *D. geminata* for obtained future study material to evaluate the effects of water contaminated by this diatom on cell models. Our preliminary results indicate the *D. geminata* altered several cell lines and cell function, and review of the literature suggested alteration in the biotic system on river with this plague. We proposed the described literature for exploring the possibility of building a model for maintaining *D. geminata* from Chilean rivers and used the biological material obtained from artificial river, like extract and contaminated water for studying the chronic effects on cells.

Keywords: didymo, cells, biotechnology

1. Introduction: threat to freshwater environment

Didymosphenia geminata (Lyngb.) M. Schmidt (*D. geminata*) has attracted considerable attention as an invasive nuisance species in natural habitats, in different country, like Chile, Argentina, Canada, Poland and New Zealand [1, 2]. The species also show nuisance effect in the southern hemisphere [3, 4]. Biological invasions are a threat to freshwater environments and the ecosystem services they provide [5, 6]. *D. geminata* is a large diatom species (~100 μm long) that produces sulphated polysaccharide stems and forms nuisance strands that can grow up to 10 cm thick with 100% coverage in streams [7]. This microalga, belonging to the brown algae diatom family, is rich in specific molecules such as antioxidants, which include polyphenols and diadinoxanthin [8]. What is the nuisance effect of this plague? The most significant impacts that this diatom presents to aquatic systems can be observed through its physical changes, substantial increases in algal biomass, stagnation of fine deposits, hydrodynamic change and their secondary effects on biogeochemical states and processes, such as redox conditions, pH and nutrient cycling in the benthic layers [9]. Recently, *D. geminata* has been molecularly identified in the rivers of southern Chile [10] and was compared with different samples; in fact, Argentinian and Chilean *D. geminata* are the same species, but the authors did not describe their origin [11]. This observation is important because the movements of *D. geminata* can improve our understanding of the species' dispersion; in fact, an essential work in New Zealand proved the invasion of *D. geminata* and gave us clues about the source of these diatoms [12].

Few studies have focused on environmental impacts, although these were mentioned indirectly in nutrient recycling mechanism and flow channel attenuation studies [13–15]. Other studies have indicated that *D. geminata* alters microenvironments and reduces fish populations [16] by altering fish community diets [17]. Also, it disturbs aquatic invertebrate communities and the filters of drinking water systems [18, 19], although it is unknown whether the contamination effect is direct. Recently, the toxic effects of microalgae on contaminated river communities [14] and the effects on *Salmo salar* spermatozoon activation time have been described [20]. Despite extensive research on *D. geminata* in recent years [21] with results on ecosystem effects, further advances on the toxicological effects are still needed.

2. Method

We used a collection of different databases, which included ScienDirect and Scopus, and we do it with a selection of the only index articles published. Based on the previous report from our group laboratory, we made a collection of the 135 different articles and review; quantitative, qualitative and mixed method evaluations, with and without comparative groups, were eligible for inclusion in the review. The metadata and abstracts from the literature search were transferred to software [EndNote X9 (Thomson Reuters, CA, USA)] and duplicated are eliminate. Two reviewers observed each title and abstract and selected studies that potentially met the eligibility criteria, or these keywords are used for exploring the internal database building whit Endnote. These papers were obtained in full, and two reviewers screened the full papers for inclusion, with any disagreements resolved by consensus with reference to the full papers and a third reviewer if necessary.

3. *Didymosphenia geminata* (Lyngb.) M. Schmidt maintenance and hydrological problems

The patterns of silica, phosphorus and electrical conductivity observed in the presence of *Cymbella* sp. and *D. geminata* have previously only been documented for *D. geminata* and phosphorus [22] which coincide with the results presented by Gretz et al. [23] in which a range of phosphorus levels in *D. geminata* has frequently been detected in the USA. The richness of microalgae does not decrease with increasing phosphorus and conductivity, so the patterns of *D. geminata* and *Cymbella* sp. are not frequently recorded for these parameters. Regarding the temperature, the results could be interpreted as an extension of the thermal niche of *Cymbella* sp. facilitated by the presence of *D. geminata* and the existence at higher temperatures than those recorded in the absence of *D. geminata*. As a hypothesis, the next process is proposed to explain the following:

1. How *D. geminata* colonises a habitat?

2. Why *D. geminata* produces a thicker mat at higher temperatures [24]?

3. How a mat of greater thickness generates a microhabitat that facilitates colonisation by *Cymbella* sp., thus increasing its thermal niche?

The development of kind of model will help to understand these issues and develop studies about the biology and relationship of *D. geminata* with other diatoms, such as *Cymbella* sp. In this case, building an artificial model, such as river model, helps us to follow this process and to suggest a biological mechanism for *D. geminata* development about other diatoms and river species.

The conditions in which *D. geminata* is established are, and the subsequent development of rock snot is even more complex. The studies have focused mainly on the physicochemical conditions of the development of *D. geminata* [1, 24], and very few investigations have incorporated other physical-hydrological factors, such as turbidity, runoff speed, shear stress, substrate [19, 25] and the like. This view leads to many degrees of freedom for assessing the main factors that condition the establishment and proliferation of *D. geminata*. It is thus necessary to reduce the degrees of freedom to model these conditions and to work in more controlled conditions through an in vitro model, such as what we intend to develop in our project.

An in vitro model of an artificial hydraulic system is designed using their hydraulic similarity (a hydraulic physical model), which maintains the development of controlled fluvial and ecological processes. The hydraulic similarity must be carried out, considering the dimensionless numbers of the hydraulics that allow for the characterisation of the regime, such as the Froude number. However, the challenge not only lies in the design via the dimensionless numbers of the hydraulic system but must also be a system that maintains the eco-hydraulic conditions that are typical of the rivers in which the establishment of *D. geminata* has been evidenced, such as the runoff speed, turbidity, depth, and type of substrate. In other words, it will be necessary to replicate as faithfully as possible the hydraulic, thermal and ecological conditions of the rivers in which the samples of *D. geminata* will be collected, but the artificial hydraulic system design must be hydraulically similar; this is the challenge of the current project being discussed.

4. *Didymosphenia geminata* (Lyngb.) M. Schmidt maintenance: the first model

We assessed whether it was possible to keep *D. geminata* in a laboratory setting close to the conditions found in a river. Accordingly, aquarium systems that regulate water variables were implemented, which allowed us to observe the maintenance or development of *D. geminata*. The substrate with the adhered *D. geminata* was collected from the rivers and placed in the aquarium. The aquarium maintained light and flow conditions and a water column of approximately 15 cm. This *D. geminata* maintenance aquarium was observed continuously for 2 months, and then a second tank was prepared with distilled water, a flow pump and uncontaminated river substrate that was washed and sterilised by an autoclave. *D. geminata* was maintained as presented in a report by Olivares et al. [20]. Notably, after 4 weeks, *D. geminata* showed a population of microalgae that appeared to exhibit cyclical behaviour, ranging from a few to large numbers of *D. geminata* per field (subjective observations). Our previous data suggested that it is possible to keep and replant *D. geminata* in artificial river conditions; we propose to improve this protocol to generate a new version for the river model [26]. This work allows us to suggest some ideas for the management of *D. geminata* in the laboratory, allowing open protocols for in vitro studies of *D. geminata* to obtain a better understanding of its biology. Observations made in this earlier study indicate that it is possible to keep *D. geminata* in the laboratory, which seems to contradict what is mentioned in the literature regarding both its maintenance [27] and preservation [28]. It may even be possible to move and transport samples from the collection sites for analysis [29] and thus achieve complex systems of general diatom cultures from biofilm communities [30].

The maintenance of *D. geminata* in the aquarium was successful following the protocol described in Section 2 and achieved a viability of 90%. In this protocol, the critically essential parameters appear to be the following: the use of polluted river substrate, a constant water flow, low temperatures and a water column of 15 cm above the substrate. Moreover, we observed a set of simple organisms, indicating a complex ecological system associated with *D. geminata*, but it is not possible to induce controlled contamination of a new complex substrate with more details in the flow; in addition, the river conditions need to be prepared and studied [20, 26].

Studies from other groups have addressed these questions by making direct observations in rivers. Observing their characteristics to model the growth of *D. geminata* [31] and evaluating the effect of phosphorus as a nutrient function or process of *D. geminata* [27, 29], the type of substrate required [19] or the effect of nutrients on the development of this diatom [32], such studies require observations in the field, which make it difficult to control environmental and seasonal variables, as well as requiring that the seasonality of the hydrologic regime be considered at the time of collection. In contrast, this project suggests a *D. geminata* maintenance system that would allow *D. geminata* studies in the laboratory and provide the basis for generating systems that could test the conditions of *D. geminata* development or control. For example, the effect of shear stress in rivers could be mimicked in scaled models using hydraulic flow to simulate runoff characteristics.

However, not all of the river model is complete, and in our experiment and those of other groups, the culture of *D. geminata* has not yet been reported. In previous work, we cited where a complex culture protocol for single *D. geminata* had explored and replicated a closed in vitro model similar to a river for the study of *D. geminata* that is not present in the existing literature. The feasibility of keeping *D. geminata* in river model suggests the possibility of replicating and maintaining it in the laboratory but does not indicate that it is a culture medium because the conditions that mimic the observed outcrop have not been determined. However, it presents a new tool for learning more about *D. geminata* and its growth conditions.

5. Cellular effects

Recent studies have suggested additional toxic impacts of *D. geminata* on invertebrates [7] and benthic river communities [18] or of the treatments used for its removal [33]. This information on the effects of the microalgae is just recently being explored. Previous data suggested that *D. geminata* can mainly affect the activation time of *Salmo salar* spermatozoa, secondary at present of the polyphenols, and a possible mechanism is the alteration of the calcium in the cells. This is an evidence of alterations in cell physiology, possibly due to the presence of polyphenols; more importantly, however, our findings indicate that the effects of this plague are more complex and should be closely watched. *D. geminata* should be considered as a toxicological agent that affects cell function.

DOI: http://dx.doi.org/10.5772/intechopen.91875

Evidence suggests an effect of water contaminated with *D. geminata* on the motility time of *Salmo salar* spermatozoa. A decrease is observed in their activation time and the percentage of motile cells, without affecting the viability of the spermatozoa or motile spermatozoon kinetics. These effects could be mediated by the polyphenol content present in *D. geminata* and could be released into rivers contaminated by this microalga. *D. geminata* has become a pollution problem in the rivers of southern Chile. It was initially detected in rivers in the "Los Lagos" Chilean area, but since then there have been at least three more regions that reportedly have this plague, and projections are that it will continue to spread if barriers are not raised. No evidence exists that it can be maintained in a cultivated system, despite its rapid growth [15].

In contrast, it is possible to find research indicating that under various conditions, it is not feasible to keep *D. geminata* in closed systems [34]. An older study suggests a prolonged protocol for keeping *D. geminata* but not that made in culture [35]. Other models have described the effects of *D. geminata* on the microflora of rivers, reporting changes mainly in aquatic species, but with no reports of similar results in more complex forms [7]. Our findings indicate that the viability of *Salmo salar* spermatozoa is not affected when exposed to water contaminated with *D. geminata*. However, the reaction time and the mass of motile spermatozoa are changed. A decrease of 50% in the activation time was shown, which alters the functions of the spermatozoa. The Powermilt compound has been used as a spermatozoon activator as described by Olivares et al. [20]. This compound produced an excellent response in our samples, with an average activation time of 4 min. To understand the sperm activation mechanism, under *D. geminata* presence, experimental observations were made about whether it was able to inhibit the Powermilt's effects on the activation times. Upon dilution with water contaminated with *D. geminata*, we observed a decreased effect of the Powermilt stock on activation; the sperm activation it was fully inhibited when using 100% *D. geminata*. This data suggests that the release of molecules into river water containing *D. geminata* would reduce the activation time of *Salmo salar* spermatozoa without affecting their kinetics. Another study with metals in the water showed effects on the kinetic parameters [36], but our results showed that the contaminated water altered the activation processes, except for the already activated spermatozoa. This outcome suggests a complex mechanism that reduces the number of cells that can be activated but not the quality of this activation, which involves a different contamination mechanism, such as metal in the water.

The organic content of *D. geminata* has proven to be rich in antioxidant polyphenols that are common in brown algae, such as diadinoxanthin. This type of molecule has cellular effects that range from beneficial to toxic, depending on the concentration. In the samples maintained or recovered from rivers, there was a ratio of approximately 200 ppm of polyphenols per 10 g of *D. geminata* that could be extracted. To explore whether it was possible to associate this organic content to the functional effects of *D. geminata*, the spermatozoa were activated with increasing concentrations of the extracts, thus showing a dose-response effect when inhibiting the time activation of the samples with an IC_{50} of 15 ppm of polyphenols. We conclude that the presence of polyphenols mediates the effect observed in the waters contaminated with *D. geminata*, but we cannot reject other mechanisms for altering cellular function.

Furthermore, in vitro cell cultures have proven to be a suitable tool for assessing the toxicity of different chemicals in fish [37]. A polyphenol study in two fish cell lines showed significant differences in the EC_{50} values for phenolic compounds (phenol and 2,4-dinitrophenol) in CHSE cells when assessing cell viability and proliferation [38].

In other study, we evaluated the cell sensibility of SHK-1 and CHSE-214 to *D. geminata*, two cell lines that will be used to assess various physiological fish farming activities (to evaluate SHK-1 and CHSE-214 cell sensitivities against *D. geminata*), which will also be used to determine multiple physiological aquacultural effects on Atlantic salmon and Chinook salmon [39, 40].

In general, the current studies on this diatom have focused on oligotrophic specimens, with no toxicity studies on species living above biota in rivers, which this project aims to elucidate for the effects of *D. geminata* on cell viability, using in vitro model of the river.

This aim is of great interest for analysing whether these same effects occur in native freshwater species, as this could be a bioindicator of water quality, as well as for other species or biological models [41]. In this regard, the reproduction of native

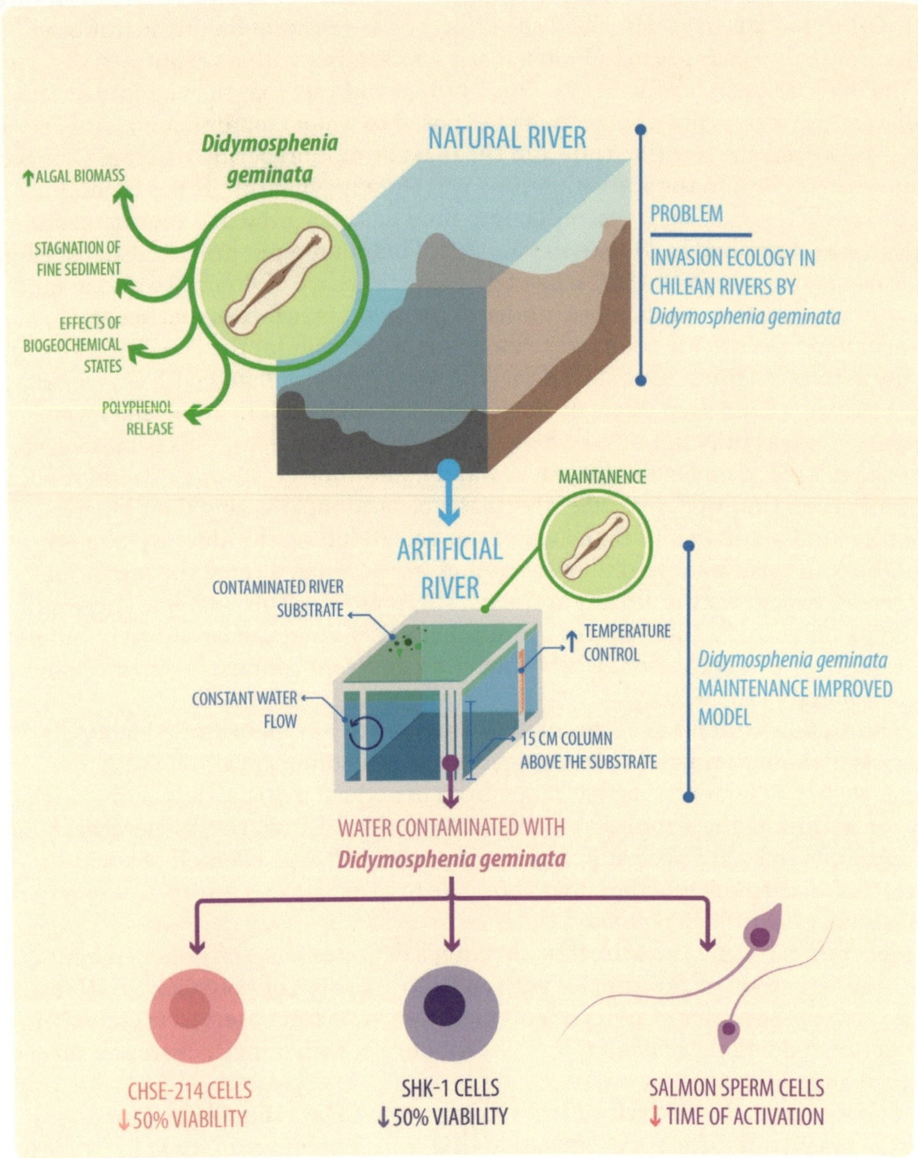

Figure 1.
A model summarizing the chapter. We present the idea of developing an in vitro *system for maintaining* D. geminata *and obtaining biological material for cell studies.*

fish species was reduced and exhibited a low success rate [42, 43]. Furthermore, there are no records where species are mentioned as being affected by the microalgae in rivers, revealing a deficiency in the research on the effects of *D. geminata* on the organs of native river species. Internationally, changes have been indicated in microinvertebrate compositions through *Salmo trutta* diet observations [17] and are consistent with the changes in the biota from rivers contaminated with *D. geminata*. Previous results suggest complex cellular-level effects of the polyphenols present in the microalgae. These effects could be considered relevant, since the compound, when concentrated and present for long periods of time, could lead to the chronic impacts (longer than 24 h), thereby generating deleterious effects on superior aquatic species. *D. geminata* can be considered toxic in upwelling conditions (blooms), where it is most active, thus altering the aquatic community's viability (macroinvertebrates and fish) because of the physical effects of river coverage and sediment trapped in estuaries [9] or the composite mass, which is mainly mucilage [44]. What we presumed is a possible development of a system of river model by maintaining *D. geminata* in a laboratory to obtain biological material to describe its cellular effects better and understand the impact of contamination by *D. geminata* in rivers. We summarise our idea in **Figure 1**, a model of the general purpose of developing a system to study *D. geminata*, and apply the study to the biological effects of rivers contaminated by this diatom. We present the models to explain the future of our work.

6. Concluding remarks

It is possible and necessary to improve a system of maintenance and study of the development of the *D. geminata* in river model. At the same time, this model helps in obtaining biological material to research at the cellular level of the effect of this contamination in rivers. Our Chapter, present a review to supports the idea at is possible the development of a system of river model and are necessary for explored and explain several mechanism of the *D. geminata* on river and different biological model, included cell model. We build the first model, but more improvement is necessary to have a more robust river model where *D. geminata* is maintained. Allowing the handling of biological samples for cell studies and this helps to us to report from this model, the controlled biological material from river model improved with *D. geminata* and without *D. geminata.* The maintenance of *D. geminata* and the effects of the organic material obtained in cellular models in scientific publications and future description of *D. geminata* biology.

Acknowledgements

We are in debt with OI 101340 and 101322 from the Universidad Mayor. Thanks to Mr Serrano from Illustrative Science for the discussion on the model.

Author details

Jorge Parodi[1*], Viviana Chavez[2,3] and Matias Peredo-Parada[3]

1 Laboratorio de Biología Celular y Molecular Aplicada, Facultad de Ciencias, Vicerrectoría de Investigación, Universidad Mayor, Sede Temuco, Temuco, Chile

2 Laboratorio de investigación Biosocial, Tonalli ltda, Temuco, Chile

3 Plataforma de Investigación en Ecohidrología y Ecohidráulica EcoHyd, Santiago, Chile

*Address all correspondence to: jparodi2010@gmail.com

References

[1] Gretz MR, Riccio ML, Kiemle SN, Domozych DS, Spaulding SA. *Didymosphenia geminata* as a nuisance diatom: Runaway stalk production results in mats with significant environmental impact. Journal of Phycology. 2007;**43**:16

[2] Blanco S, Ector L. Distribution, ecology and nuisance effects of the freshwater invasive diatom *Didymosphenia geminata* (Lyngbye) M. Schmidt: A literature review. Nova Hedwigia. 2009;**88**(3-4):347-422

[3] Kilroy C, Unwin M. The arrival and spread of the bloom-forming, freshwater diatom, *Didymosphenia geminata*, in New Zealand. Aquatic Invasions. 2011;**6**(3):249-262

[4] Reid BL, Hernandez KL, Frangopulos M, Bauer G, Lorca M, Kilroy C, et al. The invasion of the freshwater diatom *Didymosphenia geminata* in Patagonia: Prospects, strategies, and implications for biosecurity of invasive microorganisms in continental waters. Conservation Letters. 2012;**5**(6):432-440

[5] Strayer DL. Alien species in fresh waters: Ecological effects, interactions with other stressors, and prospects for the future. Freshwater Biology. 2010;**55**(s1):152-174

[6] Tomas P, Oscoz J, Duran C, Fernandez D, Marin JP. Distribution of the bloom-forming diatom *Didymosphenia geminata* in the Ebro River basin (Northeast Spain) in the period 2006-2009. Aquatic Invasions. 2010;**5**(3):285-289

[7] Richardson D, Oleksy I, Hoellein T, Arscott D, Gibson C, Root S. Habitat characteristics, temporal variability, and macroinvertebrate communities associated with a mat-forming nuisance diatom (*Didymosphenia geminata*) in Catskill mountain streams, New York. Aquatic Sciences. 2014;**76**(4):553-564

[8] Lohr M, Wilhelm C. Algae displaying the diadinoxanthin cycle also possess the violaxanthin cycle. Proceedings of the National Academy of Sciences of the United States of America. 1999;**96**(15):8784-8789

[9] Reid B, Torres R. *Didymosphenia geminata* invasion in South America: Ecosystem impacts and potential biogeochemical state change in Patagonian rivers. Acta Oecologica. 2014;**54**:101-109

[10] Jaramillo A, Osman D, Caputo L, Cardenas L. Molecular evidence of a *Didymosphenia geminata* (Bacillariophyceae) invasion in Chilean freshwater systems. Harmful Algae. 2015;**49**:117-123

[11] Ladrera R, Goma J, Prat N. Effects of *Didymosphenia geminata* massive growth on stream communities: Smaller organisms and simplified food web structure. PLoS One. 2018;**13**(3):e0193545

[12] Kilroy C, Novis P. Is *Didymosphenia geminata* an introduced species in New Zealand? Evidence from trends in water chemistry, and chloroplast DNA. Ecology and Evolution. 2018;**8**(2):904-919

[13] Ellwood NTW, Whitton BA. Importance of organic phosphate hydrolyzed in stalks of the lotic diatom *Didymosphenia geminata* and the possible impact of atmospheric and climatic changes. Hydrobiologia. 2007;**592**:121-133

[14] Kilroy C, Larned ST. Contrasting effects of low-level phosphorus and nitrogen enrichment on growth of the mat-forming alga

Didymosphenia geminata in an oligotrophic river. Freshwater Biology. 2016;**61**(9):1550-1567

[15] Sundareshwar PV, Upadhayay S, Abessa M, Honomichl S, Berdanier B, Spaulding SA, et al. *Didymosphenia geminata*: Algal blooms in oligotrophic streams and rivers. Geophysical Research Letters. 2011;**38**(10):1-5

[16] Jellyman PG, Clearwater SJ, Clayton JS, Kilroy C, Blair N, Hickey CW, et al. Controlling the invasive diatom *Didymosphenia geminata*: An ecotoxicity assessment of four potential biocides. Archives of Environmental Contamination and Toxicology. 2011;**61**(1):115-127

[17] James DA, Chipps SR. Influence of *Didymosphenia geminata* blooms on prey composition and associated diet and growth of brown trout. Transactions of the American Fisheries Society. 2016;**145**(1):195-205

[18] Gillis CA, Chalifour M. Changes in the macrobenthic community structure following the introduction of the invasive algae *Didymosphenia geminata* in the Matapedia River (Qubec, Canada). Hydrobiologia. 2010;**647**(1):63-70

[19] Bergey EA, Cooper JT, Phillips BC. Substrate characteristics affect colonization by the bloom-forming diatom *Didymosphenia geminata*. Aquatic Ecology. 2010;**44**(1):33-40

[20] Olivares P, Orellana P, Guerra G, Peredo-Parada M, Chavez V, Ramirez A, et al. Water contaminated with *Didymosphenia geminata* generates changes in *Salmo salar* spermatozoa activation times. Aquatic Toxicology. 2015;**163**:102-108

[21] Manrique JM, Uyua NM, Bauer GA, Santinelli NH, Gabriela AM, Sala SE, et al. Nuisance *Didymosphenia geminata* blooms in the Argentinean Patagonia: Status and current research trends. Aquatic Ecosystem Health & Management. 2017;**20**(4):361-368

[22] Bray J, O'Brien J, Harding JS. Production of phosphatase and extracellular stalks as adaptations to phosphorus limitation in Didymosphenia geminata (Bacillariophyceae). Hydrobiologia. 2017;**784**(1):51-63

[23] Gretz MR, Riccio ML, Kiemle SN, Domozych DS, Spaulding SA. Didymosphenia geminata as a nuisance diatom: Runaway stalk production results in mats with significant environmental impact. Journal of Phycology. 2007;43:16

[24] Bothwell ML, Taylor BW, Kilroy C. The Didymo story: The role of low dissolved phosphorus in the formation of *Didymosphenia geminata* blooms. Diatom Research. 2014;**29**(3):229-236

[25] Cullis JDS, McKnight DM, Spaulding SA. Hydrodynamic control of benthic mats of *Didymosphenia geminata* at the reach scale. Canadian Journal of Fisheries and Aquatic Sciences. 2015;**72**(6):902-914

[26] Parodi J, Olivares P, Chavez V, Peredo-Parada M. Laboratory handling of *Didymosphenia geminata* (lyngbye) schmidt and the effect of control efforts on viability. Advances in Bioscience and Biotechnology. 2015;**06**(08):508-516

[27] Bothwell ML, Kilroy C. Phosphorus limitation of the freshwater benthic diatom *Didymosphenia geminata* determined by the frequency of dividing cells. Freshwater Biology. 2011;**56**(3):565-578

[28] Agrawal SC, Singh V. Viability of dried filaments, survivability and reproduction under water stress, and survivability following heat and UV exposure in *Lyngbya martensiana*,

Oscillatoria agardhii, *Nostoc calcicola*, *Hormidium fluitans*, *Spirogyra* sp. and *Vaucheria geminata*. Folia Microbiologica. 2002;**47**(1):61-67

[29] Aboal M, Marco S, Chaves E, Mulero I, Garcia-Ayala A. Ultrastructure and function of stalks of the diatom *Didymosphenia geminata*. Hydrobiologia. 2012;**695**(1):17-24

[30] Debenest T, Silvestre J, Coste M, Delmas F, Pinelli E. A new cell primo-culture method for freshwater benthic diatom communities. Journal of Applied Phycology. 2008;**21**(1):65

[31] Cullis JDS, Gillis CA, Bothwell ML, Kilroy C, Packman A, Hassan M. A conceptual model for the blooming behavior and persistence of the benthic mat-forming diatom *Didymosphenia geminata* in oligotrophic streams. Journal of Geophysical Research: Biogeosciences. 2012;**117**(2):1-11

[32] Shimada C, Nakamachi M, Tanaka Y, Yamasaki M, Kuwata A. Effects of nutrients on diatom skeletal silicification: Evidence from *Neodenticula seminae* culture experiments and morphometric analysis. Marine Micropaleontology. 2009;**73**(3):164-177

[33] Larned ST, Kilroy C. Effects of *Didymosphenia geminata* removal on river macroinvertebrate communities. Journal of Freshwater Ecology. 2014;**29**(3):345-362

[34] Czarnecki DB. The loras college non-diatom fresh-water algal culture collection. Cryptogamie Algologie. 1988;**9**(3):203-209

[35] Kataoka H. Phototropic responses of *Vaucheria geminata* to intermittent blue light stimuli. Plant Physiology. 1979;**63**(6):1107-1110

[36] Dietrich GJ, Dietrich M, Kowalski RK, Dobosz S, Karol H, Demianowicz W, et al. Exposure of rainbow trout milt to mercury and cadmium alters sperm motility parameters and reproductive success. Aquatic Toxicology. 2010;**97**(4):277-284

[37] Bols N, Dixon D, Hodson P, Kaiser K. The use of cell cultures as an indicator of contaminant toxicity to fish. Aquatic Toxicology. 1985;**6**:147-155

[38] Castaño A, Tarazona J. Acute toxicity of selected metals and phenols on RTG-2 and CHSE-214 fish cell lines. Bulletin of Environmental Contamination and Toxicology. 1995;**55**:222-229

[39] Lee P, Zou J, Holland J, Martin S, Scott C, Kanellos T, et al. Functional characterisation of a TLR accessory protein, UNC93B1, in Atlantic salmon (*Salmo salar*). Developmental and Comparative Immunology. 2015;**50**:38-48

[40] Srikanth K, Pereira E, Duarte A, Rao J. Evaluation of cytotoxicity, morphological alterations and oxidative stress in Chinook salmon cells exposed to copper oxide nanoparticles. Protoplasma. 2016;**253**(3):873-884

[41] Kawecka B, Sanecki J. *Didymosphenia geminata* in running waters of southern Poland—Symptoms of change in water quality? Hydrobiologia. 2003;**495**(1-3):193-201

[42] Cifuentes R, González J, Montoya G, Jara A, Ortíz N, Piedra P, et al. Relación longitud-peso y factor de condición de los peces nativos del río San Pedro (cuenca del río Valdivia, Chile). Gayana (Concepción). 2012;**76**(Suppl 1):86-100

[43] Montoya G, Jara A, Solis-Lufí K, Colin N, Habit E. Primeros estadios del ciclo de vida de peces nativos del Río San Pedro (Cuenca del Rio Valdivia, Chile). Gayana (Concepción). 2012;**76**:86-100

[44] Wysokowski M, Bartczak P, Zoltowska-Aksamitowska S, Chudzinska A, Piasecki A, Langer E, et al. Adhesive stalks of diatom *Didymosphenia geminata* as a novel biological adsorbent for hazardous metals removal. Clean-Soil Air Water. 2017;**45**(11):1600678

Section 2

Sustainable Aquaculture

Chapter 7

Site Suitability Analysis of Infrastructure Facilities for Giant Freshwater Prawn Farming

Benjamin Ezekiel Bwadi and Firuza Begham Mustafa

Abstract

Infrastructure facilities play important roles in any aquaculture business. This study is to assess the significance of infrastructure facilities in the suitability of site for giant freshwater prawn farming in Negeri, Malaysia. Some of the infrastructure facilities that determine the suitability of a location for prawn farming include road, electricity, market, and availability of hatcheries. Infrastructure facilities data were collected from various institutions for the analysis. Geographic information system (GIS) was used to determine the appropriate area for prawn farming. The result shows that 496,198.75 ha was most suitable, 105,414.82 ha was moderately suitable, and 63,733.73 ha was regarded not suitable. It was further revealed by the study that the infrastructure facilities of the study area have great potential for prawn farming, but the sources of fry serve as a limiting factor. It is recommended that more infrastructural facilities such hatcheries, rural road construction, and electric power supply could be established to facilitate prawn farming in the study area.

Keywords: aquaculture, analytic hierarchy process, giant freshwater prawn, infrastructure facilities

1. Introduction

A successful development of any aquaculture activity and freshwater prawn production in particular does not only require a suitable natural environment and technical method but also adequate and appropriate infrastructure facilities. Significant infrastructure facilities needed may include good and adequate road network system, market, electricity supply, and fry sources or hatchery [1].

Infrastructure facilities influence the production of prawn and need to be taken into consideration in fostering and planning for farming it. Even though the environmental and biological conditions are favourable for the production of prawn, it may fail if the infrastructure facilities are not favourable [2]. Thus, the infrastructure improvement of prawn farming in the area is a major objective for developing and sustaining prawn farming. However, there was limited study in the Negeri Sembilan area on the significance of infrastructure facilities for the suitability of site for prawn farming. The aim of this study is to assess the significance of infrastructure facilities in the suitability of site for giant freshwater prawn farming in Negeri.

2. Methods

2.1 The study area

Negeri Sembilan is located between 2°.43′ 54.5268″ N latitude and 102°.15′ 9.0072″ E longitude, occupying about 6645 km^2 in Malaysia with the population of about 1.7 million people in 2012. It borders Selangor at the northeast, Pahang at the north, Johor at the east, and Melaka at the south with the capital in Seremban. The people are engaged in farming paddy rice at the valley of the steep hills, rubber trees are extensively cultivated, and oil palms are grown. Prawn farming is one of the major livelihoods for the inhabitant. The State contributed about 22% of prawn production in Malaysia in 2014 [3].

2.2 Factors and class weights (standardisation)

The actual factors and class weights of the parameters involved in the study are needed to generate the land suitability map. The AHP was systematically used to determined theses. The AHP criteria were developed based on the expert's survey interviews. The experts were asked to determine the relative importance of each factor. The process assesses the relative importance of all the parameters by allocating weights for each of them in the hierarchy order, and the suitability weight for each class of the factors was assigned in the last level of the hierarchy. Usually, the priority of each factor involved in the AHP analysis is calculated based mainly on the opinions of the experts [4, 5]. Prioritisation is the determination of the relative significance of the criteria which needs brain stomping among experts to assign values on a Saaty's nine-point scale [6] for a pairwise comparison of criteria.

The pairwise comparison matrix was applied to determine the weight and consistency of each criterion at each level of hierarchy by relative rating. The nine-point rating scale was used where 1 represents equal importance (i.e., two factors contributing equally to the objective), 3 represents moderate importance (one factor slightly favoured over another), 5 indicates strongly important, 7 stands for very strongly important, and 9 stands for extreme importance (as earlier mentioned in Chapter 3 of this study). It is a score systematically indicating the relative rating from most important (9–1) or the least important (1/9, 1/8, 1/7… ..1/2, 1). The AHP results in combination with information collected from other methods were used to describe the land suitability analysis for prawn farm identification, as well as the opportunities and challenges for prawn farming in the current site.

The AHP as a multi-criteria evaluation method was applied to determine the weight of each criterion. The principle behind AHP is in the construction of a three-level hierarchy model with the goal, the criteria (objectives), and the sub-criteria (attributes) which are at the bottom layer of the hierarchy [7]. Inputs of experts are considered as the pairwise comparison, and the best criteria will be selected according to the highest rank between the criteria.

Multi-criteria evaluation is a process that incorporates multiple and conflicting criteria, which allow solving a wide range of complex problems and transforming them into decision-making. The AHP developed by Saaty [8] for doing pairwise comparison matrix is a tool required for comparing alternatives with respect to a set of criteria. The criteria were ranked according to the order of importance. Some relative weights were assigned to the criteria indicating the degree of importance or preference of each criterion with respect to the other criterion.

Expert opinion was usually required to rank the criteria by assigning a score to each criterion [9]. For this study 30 experts were drawn from the Department of

Fisheries (DoF) who were knowledgeable in the field of prawn farming and have relevant information to assign weight to each criterion [10, 11]. However, out of the 30 experts, only 20 were found consistent. These pairwise comparisons were then applied as the input to generate a ratio matrix, and the relative weights are created as the output [12].

The AHP pairwise comparison matrix calculates the weight for each criterion and factor (wi) by taking the eigenvector corresponding to the largest eigenvalue of the matrix and normalising the sum of the component to 1 as expressed below:

$$\sum_{i=1}^{n} wi = 1 \tag{1}$$

Then the final importance of each criterion was calculated. The main input is the pairwise comparison matrix 'A' of n criteria proposed by Saaty's scale, in the order of (n x n) as described in Eq. (2):

$$\mathrm{A} = [\mathrm{aij}], \mathrm{ij} = 1, 2, 3 \ldots \mathrm{n} \tag{2}$$

where A is a matrix with elements aij. The matrix usually has a reciprocity defined as.

$$\mathrm{aij} = 1/\mathrm{aji}\ (n(n-1)/2) \tag{3}$$

After creating this matrix, it is then normalised as matrix B:

$$\mathrm{B} = [\mathrm{Bij}], \mathrm{ij} = 1, 2, 3 \ldots \mathrm{n} \tag{4}$$

in which B is the normalised matrix of A with the elements bij expressed as

$$bij = \frac{\mathrm{aij}}{\sum_{i=1}^{n} aij = 1, 2, 3, \ldots n} \tag{5}$$

Each weight value is computed as shown below:

$$wi = \frac{\sum_{i=1}^{n} bij}{\sum_{i=1}^{n} \sum_{i=1}^{n} bij}, i, j = 1, 2, 3 \ldots, n \tag{6}$$

A mistake may be made in preference during the survey stage [13]. Therefore, Saaty [14] introduced a single mathematical index to make sure that the pairwise comparison matrix is consistent by applying the consistency ratio (CR). The consistency ratio is computed by dividing the *CI* by the *RI*. The equation is expressed as follows:

$$CR = \frac{CI}{RI} \tag{7}$$

where (*CR*) = consistency ratio; (*CI*) = consistency index; (*RI*) = random index (mean value) depending on the computed matrix order set by Saaty [8].

To ensure the reliability of the relative importance applied, the AHP provides a certain measure to determine inconsistency of the judgments. Based on the priorities of the reciprocal matrices, the consistency ratio can be calculated. CR < 0.1 indicates that the level of consistency in the pairwise comparison is acceptable. But if CR > 0.10 it means that there is inconsistency in the evaluation process and the

n	1	2	3	4	5	6	7	8	9	10
RI	0	0	0.58	0.9	1.12	1.24	1.32	1.41	1.45	1.49

Table 1.
Table of random index (RI).

process needs to be recomputed or else the AHP may not yield meaningful results [6].

Consistency ratio simplifies the assessment of possible events and measures logical inconsistencies of the decision-maker and judgement [15]. It denotes the probability where the matrix judgments were randomly formed [16].

The consistency ratio depends upon the eigenvector (λ_{Max}) and the consistency index (CI). Therefore, one needs to find the vector w of the order n such that $A \times w = \lambda \times w$.

where w is the eigenvector (i.e., weight vector) and λ is the eigenvalue.

where $\lambda_{max} \times w \geq n$ and λ_{max} is the principal eigenvalue of the matrix. Therefore, the inconsistency of the judgement is reflected in the differences between λ_{max} and n. The process computes a consistency index (CI) to check the consistency of the pairwise comparison matrix:

$$\mathrm{CI} = \frac{\lambda_{\max} - n}{n - 1} \quad (8)$$

where (CI) refers to the consistency index, (n) refers to the number of criteria used (**Table 1**), and (λ_{Max}) refers to the average value of the consistency vector (the highest eigenvector).

3. Results and discussion

The AHP pairwise comparison matrix results show as follows: The consistency ratio for the infrastructure facilities factors for site suitability for prawn farming was 0.0068. This value demonstrated that the comparison of criteria was consistent. The result show that distance to electricity criteria ranked first with 59.2%, followed by distance to roads with 21.7%, distance to market with 12.6%, and distance to fry source with 6.4% as shown in **Table 2**. According to the key informants (experts), infrastructure facilities play a key role in prawn farming, given good water quality and soil characteristics which are environmental factors; without the infrastructure

Infrastructure facilities	Pairwise comparison matrix				Normalisation and the computation of infrastructure facilities criteria weight							
	[1]	[2]	[3]	[4]	[1]	[2]	[3]	[4]	Sum	Eigenvector (weight)	Weightage (%)	Ranking
[1] Road	1	1/4	3	3	0.18	0.16	0.32	0.21	0.87	0.22	21.7	2
[2] Electricity	4	1	5	7	0.70	0.62	0.54	0.50	2.36	0.59	59.2	1
[3] Market	1/3	1/5	1	3	0.06	0.13	0.10	0.21	0.50	0.13	12.6	3
[4] Fry	1/3	1/7	1/3	1	0.06	0.09	0.04	0.08	0.27	0.06	6.4	4
Total	**5.67**	**1.59**	**9.33**	**14.00**	**1.00**	**1.00**	**1.00**	**1.00**		**1.00**	**100%**	

Table 2.
Pairwise comparative matrix for evaluating the infrastructure facilities for prawn farming.

facilities, prawn farming will not be successful [17]. Major differences in the factors ranking for giant freshwater prawn farming were the relative importance placed on infrastructure where they were higher weights in relation to water and soil as infrastructure plays a vital role in farming in the study area.

3.1 Infrastructural facilities criteria map for prawn farming

The infrastructural facilities aspects of the study comprised of distance to road, distance to market, distance to electricity, distance to fry source, or hatchery layers which were overlay in the GIS environment to generate the overall infrastructure facilities map (**Figure 1**).

Prawn farming operations were affected by infrastructural factors [18]. One of the requirements for successful prawn farming is a good road network. Foods and other necessary equipment are transported to the farm and market. Therefore, prawn farms should be close to the road for easy and quick access. The distance to road suitability classification showed that 90.5% of Negeri Sembilan was close to access roads within less than 2 km. Road accessibility was limited in the hilly and dense forested area where roads construction was difficult.

Electricity is a vital factor in determining the success of prawn farming due to the power supply to power the farm's machines for prawn production [19]. Any area lying within less than 3 km to the main electric power supply line are considered the most suitable site for prawn farming. Major roads and cities coincided with areas with good electricity supply in the study area. The area between 7 and 12 km was considered moderately suitable. In this study, more than half of the area have electricity distribution within the suitable range. The not suitable area falls within areas greater than 12 km away from the main source of electric power line occupying 0.4% of the study area at the extreme north of the study area.

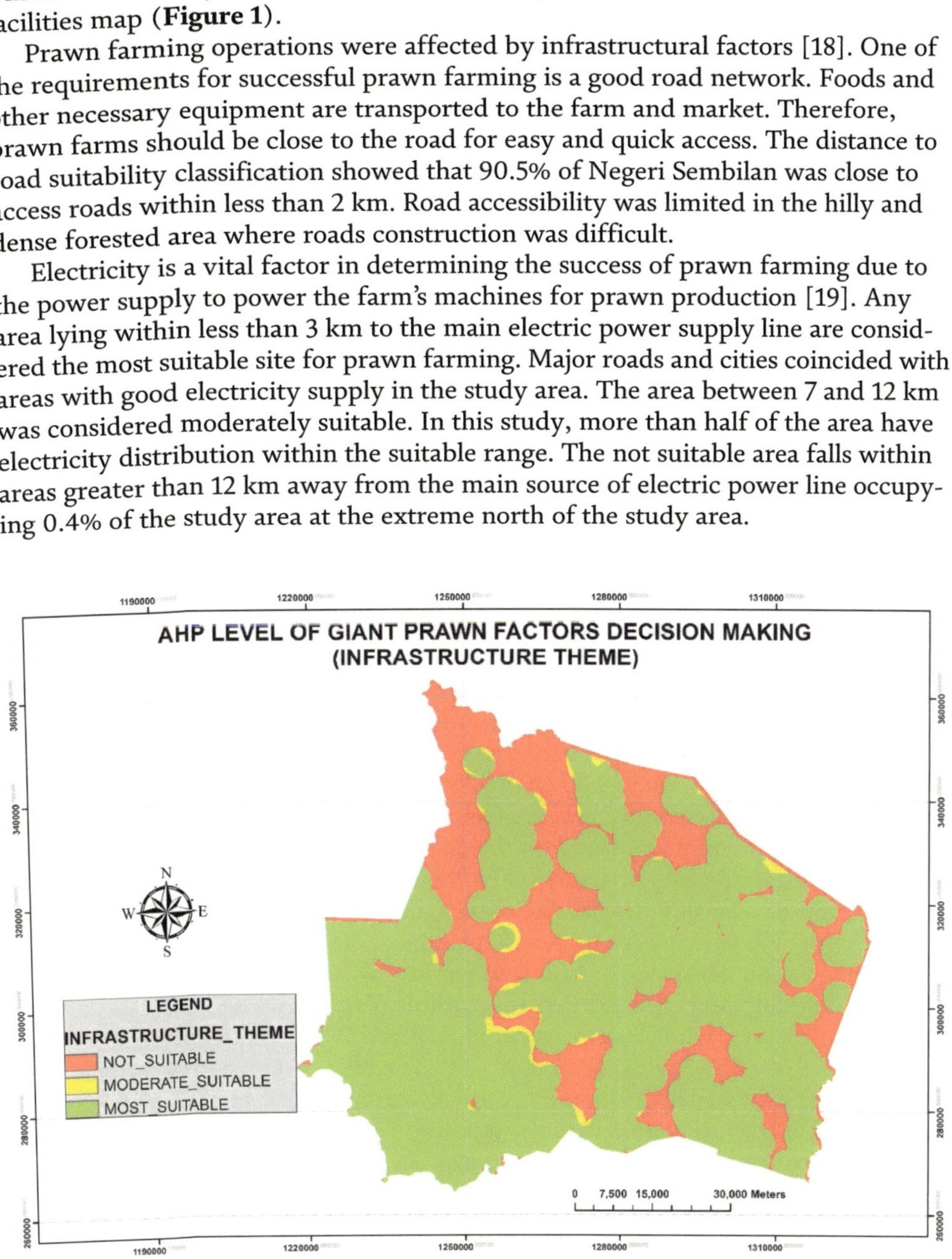

Figure 1.
The infrastructure facilities criteria suitability map for prawn farming in Negeri Sembilan,

No.	Infrastructure layer	Area (Ha)	Percentage
1	Most suitable	496,198.75	75
2	Moderate suitable	105,414.82	16
3	Not suitable	63,733.73	10
	Total area	665,347.31	100

Table 3.
Infrastructure criteria suitable area for prawn farming.

Demand for prawn is very important. Hence, prawn farm should be sited where there is high demand for the product. Quality of prawn deteriorates with time after harvest, and so it requires a prompt and ready market. Therefore, the proximity of farm to the market is very important.

Source of fries to the farm is also very important to the establishment of the farm. Fry mortality increases with long-distance travel as they are susceptible to bad environmental condition; therefore, it is advisable that hatcheries be located nearer the farm. It was discovered from the study that most hatcheries as sources of fries were located some kilometres far away from the study area such as Perak and Kedah where the farmers must travel to purchase fries which takes about 3–4 h to reach and another 3–4 h to come back, which was considered not appropriate for prawn farming. This increases the mortality rate of the fries (juvenile) and cost of production in terms of transportation. According to New [20], prawn farms should be located close to a hatchery because the mortality of the fry increases with long-distance travel to obtain it. From the study, it was found that only a few hatcheries were established in the study area.

The southwestern area of Negeri Sembilan is better equipped in terms of the infrastructural facilities. The most suitable area was basically regions along the major roads linking Negeri Sembilan and Selangor state as one move to the Kuala Lumpur, Seremban the state capital region, and the major towns. Electrification exists almost beside the roads which are significant to prawn farming. The unsuitable areas are in the northern parts with pockets of most suitable around Jelebu. The infrastructure facilities result shown in **Table 3** indicates that 75% (496,198.75 ha) of the area for prawn farming are most suitable, 16% (105,414.82 ha) as moderately suitable, and 10% (63,733.73 ha) as not suitable. This confirmed the study conducted by Olaniyi et al. [21] that good infrastructure determines the productivity of prawn of an area.

The market potential was based on the distance of the farm to the settlements and the population density of an area. The Negeri Sembilan region was largely good in terms of market potential due to the population density of 159 persons per km^2 with high demand for prawn in the region. Seremban District with the highest population density of 586 persons per km^2 was the most suitable market potential, Port Dickson with 191 persons per km^2, Tampin with 99 persons per km^2, Jempol District with 79 persons per km^2, Kuala Pilah with 64 persons per km^2, and Jelebu with 29 persons per km^2 [22]. The relative location of the area near Kuala Lumpur, the Federal capital territory, also provides the area with good market potential as prawns were sold to the restaurant for local consumption. The farms were located relatively near the market areas. The Seremban District with the population of about 555,935 (2012 census) with the land area 397,185 km and Port Dickson provide the most suitable area [22]. The rest of the districts were moderately suitable with pockets of the most suitable area. The extreme north of Jelebu District with less population was considered not suitable for a market potential.

4. Conclusion

The study integrates the GIS and AHP methods to evaluate the infrastructure facility of an area and its suitability for prawn farming. The approaches were able to identify areas with appropriate infrastructures for the sustainable development of prawn farming in Negeri Sembilan. The factor layers considered included distance to road, distance to market, distance to electricity, and distance to fry source.

From the result of the analysis, infrastructure facilities have been identified as major factors that have great influence on prawn farming. This confirms the study by Hossain and Das [1] who identified transportation facilities, electricity, market, and availability of hatcheries as some infrastructure facilities having significant influence on prawn farming.

The infrastructure facilities suitability map generated will be useful to farmers, stakeholders, policy-makers and decision-makers to improve the efficiency of land use by selecting which site is appropriate to farm prawn. They will lead to increased yield and the optimal utilisation of the available land resources. This can reduce the overall cost of production and conflicts among land users. Infrastructural facility developments in the area such as establishment of hatcheries, rural road construction, and electric power supply to facilitate prawn farming are recommended.

Author details

Benjamin Ezekiel Bwadi[1*] and Firuza Begham Mustafa[2*]

1 Department of Geography, Taraba State University, Jalingo, Nigeria

2 Department of Geography, University of Malaya, Malaysia

*Address all correspondence to: bwadiben@gmail.com and firuza@um.edu.my

References

[1] Hossain MS, Das NG. GIS-based multi-criteria evaluation to land suitability modelling for giant prawn (*Macrobrachium rosenbergii*) farming in Companigonj Upazila of Noakhali, Bangladesh. Computers and Electronics in Agriculture. 2010;**70**(1):172-186

[2] Ahmed N. Socio-economic aspects of freshwater prawn culture development in Bangladesh. 2001

[3] DoF. Annual Fisheries Reports of Malaysia, 2014. Ministry of Agriculture and Agro-allied Industries (MoA): Department of Fisheries; 2016

[4] Sindhu S, Nehra V, Luthra S. Solar energy deployment for sustainable future of India: Hybrid SWOC-AHP analysis. Renewable and Sustainable Energy Reviews. 2017;**72**:1138-1151

[5] Vaidya OS, Kumar S. Analytic hierarchy process: An overview of applications. European Journal of Operational Research. 2006;**169**(1):1-29

[6] Saaty TL. The analytic hierarchy process: Planning. In: Priority Setting. Resource Allocation. MacGraw-Hill, New York International Book Company; 1980. p. 287

[7] Dehe B, Bamford D. Development, test and comparison of two multiple criteria decision analysis (MCDA) models: A case of healthcare infrastructure location. Expert Systems with Applications. 2015;**42**(19): 6717-6727

[8] Saaty TL. A scaling method for priorities in hierarchical structures. Journal of Mathematical Psychology. 1977;**15**(3):234-281

[9] Nguyen TT, Verdoodt A, Van YT, Delbecque N, Tran TC, Van Ranst E. Design of a GIS and multi-criteria based land evaluation procedure for sustainable land-use planning at the regional level. Agriculture, Ecosystems & Environment. 2015;**200**:1-11

[10] Capraz O, Meran C, Wörner W, Gungor A. Using AHP and TOPSIS to evaluate welding processes for manufacturing plain carbon stainless steel storage tank. Archives of Materials Science. 2015;**158**:158

[11] García JL, Alvarado A, Blanco J, Jiménez E, Maldonado AA, Cortés G. Multi-attribute evaluation and selection of sites for agricultural product warehouses based on an analytic hierarchy process. Computers and Electronics in Agriculture. 2014;**100**: 60-69

[12] Kumar V, Jain K. Site suitability evaluation for urban development using remote sensing, GIS and analytic hierarchy process (AHP). In: Paper presented at the Proceedings of International Conference on Computer Vision and Image Processing. 2017

[13] Bagheri M, Sulaiman W, Vaghefi N. Land use suitability analysis using multi criteria decision analysis method for coastal management and planning: A case study of Malaysia. Journal of Environmental Science and Technology. 2012;**5**(5):364

[14] Saaty TL. Decision making—The analytic hierarchy and network processes (AHP/ANP). Journal of Systems Science and Systems Engineering. 2004;**13**(1):1-35

[15] Pramanik MK. Site suitability analysis for agricultural land use of Darjeeling district using AHP and GIS techniques. Modeling Earth Systems and Environment. 2016;**2**(2):1-22

[16] Park S, Jeon S, Kim S, Choi C. Prediction and comparison of urban growth by land suitability index

mapping using GIS and RS in South Korea. Landscape and Urban Planning. 2011;**99**(2):104-114

[17] Ahmed N, Allison EH, Muir JF. Using the sustainable livelihoods framework to identify constraints and opportunities to the development of freshwater prawn farming in southwest Bangladesh. Journal of the World Aquaculture Society. 2008;**39**(5): 598-611

[18] Nath SS, Bolte JP, Ross LG, Aguilar-Manjarrez J. Applications of geographical information systems (GIS) for spatial decision support in aquaculture. Aquacultural Engineering. 2000;**23**(1):233-278

[19] New MB. Farming Freshwater Prawns: A Manual for the Culture of the Giant River Prawn (*Macrobrachium rosenbergii*). Food & Agriculture Organization; 2002

[20] New MB. Freshwater Prawn Culture: The Farming of *Macrobrachium rosenbergii*. John Wiley & Sons; 2008

[21] Olaniyi A, Ajiboye A, Abdullah A, Ramli M, Sood A. Agricultural land use suitability assessment in Malaysia. Bulgarian Journal of Agricultural Science. 2015;**21**(3):560-572

[22] DoSM. Department of Statistics Malaysia, Official Portal. Negeri Sembilan @ a Glance. 2017

Chapter 8

Floating Cage: A New Innovation of Seaweed Culture

Ma'ruf Kasim, Abdul Muis Balubi, Ahmad Mustafa, Rahman Nurdin, Rahmad Sofyan Patadjai and Wardha Jalil

Abstract

Eucheumatoid cultivation continues to expand with a variety of methods that can increase production. This chapter will discuss an innovation in seaweed cultivation of the genus *Eucheuma*, which is the prime marine commodity in the tropical regions of the world. Research conducted during 2015-2017 and 2019 in Southeast Sulawesi Province, Indonesia, provided an overview of the use of floating cage that showed very significant growth results. The research result showed that the growth rates of *Eucheuma denticulatum* and *Kappaphycus alvarezii* in floating cage seemed faster and resulted in better thallus morphology. Daily production of *E. denticulatum* and *K. alvarezii* that were cultivated in floating cage was higher than daily production of *E. denticulatum* and *K. alvarezii* cultivated on longline. Specific growth rate (SGR) of *E. denticulatum* and *K. alvarezii* cultivated by using floating cage method was also higher than *E. denticulatum* and *K. alvarezii* cultivated by using longline method. Moreover, the cultivation by using floating cages produces good growth rates with no effect of herbivore attacks.

Keywords: *Kappaphycus alvarezii*, *Eucheuma denticulatum*, floating cage, longline

1. Introduction

Seaweed cultivation is a good activity for improving the welfare of coastal communities. Several countries in Southeast Asia such as in the Philippines [1] cultivate seaweed with a fairly high production. *Kappaphycus alvarezii* is a type of macroalgae that is widely cultivated by the community. The morphological form of this species looks very bright and has many branches. This species has large primary thallus, and the color of the thallus is darker than that of thallus branches that look bright from the maintenance of vegetative branches [2]. Old branches of *Eucheuma denticulatum* that has been cultivated for more than 4 weeks are seen to spread. This distribution is thought to be a major factor causing this species to grow on coral reefs in Kaneohe Bay, Hawaii. Meanwhile, some of the most widely cultivated species are *K. alvarezii*, *Eucheuma striatum*, and *E. denticulatum* [2]. *E. denticulatum* has been widely cultivated in several countries such as the Philippines since 1971 [3–5]. The production of *Eucheuma* species continued to increase very rapidly in the initial cultivation process in 1971 with initial seedlings of less than 1000 Mt. of dry weight up to 100,000 Mt. of dry weight [3]. Two countries in Southeast Asia which intensively cultivate *E. cottonii* species are the Philippines and Indonesia. Specifically in Indonesia, until 1990, this seaweed production reached 6 Mt. per year of cultivation area [6]. Seaweed species of *Eucheuma* sp. is a group of lower plant known as seaweed and has a fairly rapid growth in nature [1]. Philippines is a country that has succeeded in developing a generative regeneration system, making the development of seaweed *Eucheuma* sp. quite fast [7]. Seedling growth tends to be rapid with changes in weights that continue to increase every week [8]. Meanwhile, Fiji Island has begun developing *K. alvarezii* aquaculture in their surrounding waters since 1970. Cultivation K. alvarezii are simple with poles and ropes and then develop to date using the longline method [9, 10]. Tuvalu has begun developing the cultivation of seaweed since 1977 [11], while Malaysia has developed cultivation since 1978 [4]. In Indonesia, the cultivation has begun to develop since 1985 [12, 13], while in Maldives, it has begun to develop in 1986 [14]. Furthermore, India and Tanzania have begun developing cultivation since 1989 [15–17], and several countries such as Vietnam, Brazil, and Venezuela have begun developing during the 1990s [18–20].

Cultivation of *K. alvarezii* developed quite well and provided a fairly good growth related to various physical chemical environmental factors that existed especially in the Yucatan region of Mexico [21]. Although well developed, growth of *K. alvarezii* species fluctuates seasonally especially in the northwestern coast of India [22]. The production value of the *K. alvarezii* species also depends on the cultivation method used. A closed method protecting *K. alvarezii* from various herbivorous attacks will be better than open methods. This happened in the waters of the south coast of Rio de Jeneiro, Brazil, illustrating that cultivation techniques with tubular netting are better than with the longline method [23].

Currently, various problems continue to afflict cultivation activities carried out by the community. One thing that really stands out from the various problems that exist is the low quality and low production of seaweed. There has been a decline in the quality of seaweed after 20 years of seaweed production in the Pacific region. Specifically in Indonesia, the low production of seaweed cultivation is caused by low quality of seeds, poor cultivation methods, intensity of attacks of pests and diseases, and marine environment that is impacted by global climate change [24]. Increase in population of herbivorous fish in the seaweed cultivation areas around the Mediterranean waters correlated with the species of epiphyte that existed. Current phenomenon that some unprotected cultivation sites provide food opportunities for herbivorous fish [25]. Recently, many studies have been done, but there are no appropriate actions and solutions to overcome the existing problems. Each stage of seaweed cultivation has its own specific problems. For this reason, it is very important to conduct a thorough and integrated research to approach every problem, starting from getting good quality seeds, good maintenance patterns, and biological, physical, chemical, and ecological factors related to seaweed resistance to the economic value of each production results that affect the selling value of seaweed.

2. Current problems in seaweed cultivation

The phenomenon of grazing by herbivorous fish is very common in all locations of seaweed cultivation (**Figure 1**). In India, especially in Krusadai Island, the production of seaweed farming of *Kappaphycus alvarezii* and *Eucheuma denticulatum* species has decreased around 10% of their growth due to attacks of herbivorous fish such as *Siganus javus* (rabbitfish), *Acanthurus* sp. (surgeonfish), *Cetoscarus* sp. (parrot fish), and sea urchin *Tripneustes* sp. [25]. Therefore, cultivation of *Kappaphycus alvarezii* in a cage can prevent seaweed from being attacked by herbivore. Cultivation with this protection system can be done with modifications in accordance with the topography of the area, and this has been done in several cultivation areas in the Philippines [26].

Figure 1.
*Grazing activity of herbivorous fish (*Siganus *sp.) on* Kappaphycus alvarezii.

The use of cage for *K. alvarezii* cultivation can be carried out in areas that have high intensity of herbivorous fish attack. However, the durability of the tools to support cultivation needs to be considered. *K. alvarezii* cultivation activity together with oysters and snapper cultivation activities is an alternative cultivation with a good confinement system [27]. Seaweed *Gracilaria gracilis* can be cultivated in a closed container to avoid various attacks by herbivore such as isopods. These animals are seen to not exist in closed cultivation conditions, thus preventing these animals from entering to eat seaweed. Herbivorous fish are not detected when this type of seaweed is treated with freshwater in an enlargement tank, which does not cause damage [28].

3. New innovation method in seaweed cultivation

Floating cage is a technological design that functions as a seaweed cultivation tool. This cultivation tool can protect seaweed from pests while reducing the potential for ice–ice disease [16, 29, 30]. Various sizes and shapes were developed as tools that can be used for seaweed cultivation. The basic materials of floating cages include PVC pipes, nets equipped with buoys as markers, and weights attached to the raft to facilitate the laying of the raft in the desired area. The shape of the tool is designed like a rectangular box with varied sizes. The surface of the raft is left

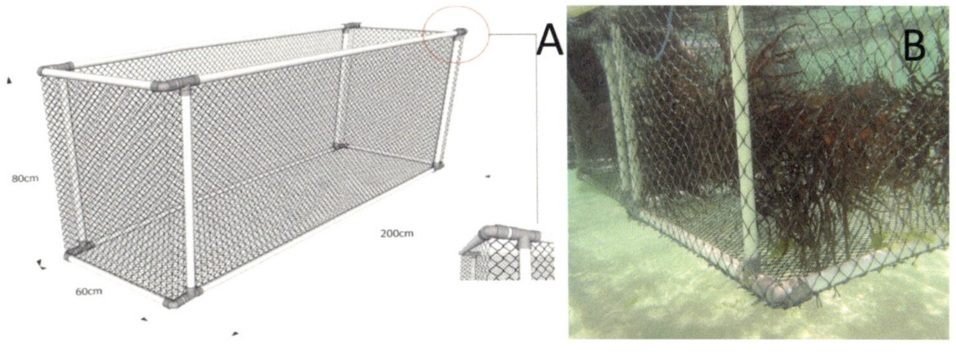

Figure 2.
Illustration of floating cage that was used during research (a). Thallus of K. alvarezii *inside the floating cage (B).*

open, while the bottom and all sides are covered using a multifilament net with a mesh size of 1 cm [31]. The process of seaweed cultivation in floating cages is done by spreading seaweed directly in floating cage without binding. In the harvesting process, floating cage can be directly drawn to the shallower parts of the coast so that the harvesting process can be more easily done (**Figure 2**).

4. Growth of thalli of *Kappaphycus alvarezii* in floating cage

The growth of thalli of *Kappaphycus alvarezii* cultivated in Lakeba Beach, Southeast Sulawesi Province, Indonesia, was different from *E. denticulatum* which was maintained by a floating net (**Figure 2**) and longline (**Figure 3**). This difference occurred during 20, 30, and 40 days of the maintenance period. More differences were caused by the shape and number of thalli that grew and the presence of disruption from pests. This was because the thalli maintained by the longline method looked new and sporadic, and the tip of each thallus appeared to be blunt. This indicated the presence of pests that ate seaweed. Meanwhile, those maintained with

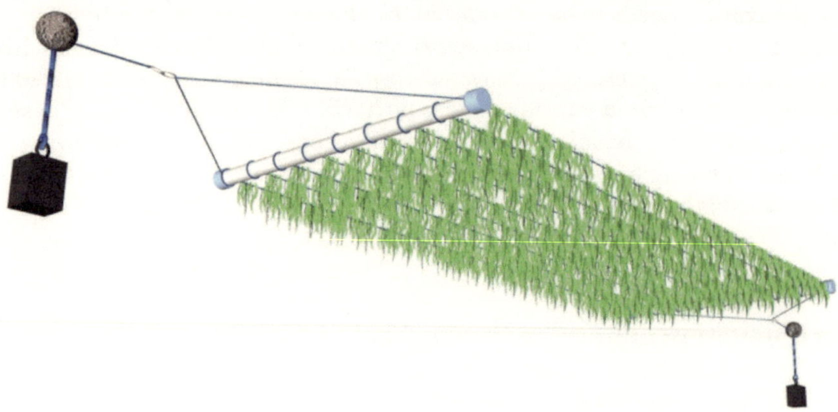

Figure 3.
Illustration of seaweed cultivated on longline.

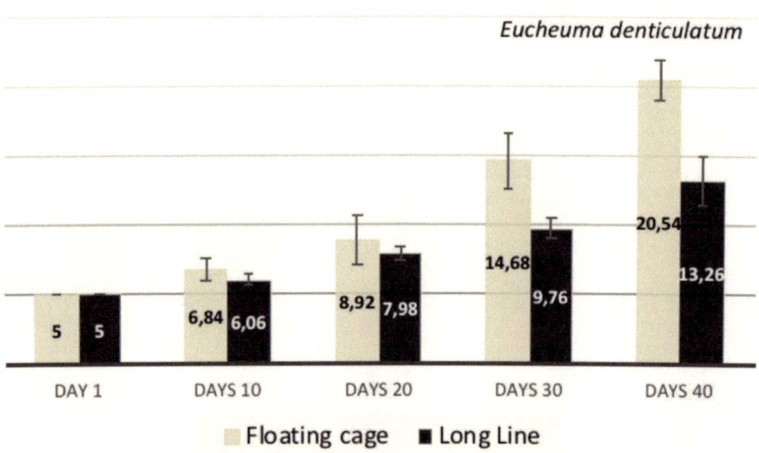

Figure 4.
Total growth of seaweed Eucheuma denticulatum *cultivated with floating cage and longline at initial weight of 5 kg.*

floating net rafts appeared to have lush thallus with tapered thallus tip. The results show that seaweed production cultivated by floating cage looked much better than that cultivated by longline. Growths of *E. denticulatum* that was cultivated using floating cage and longline looked different. The growth of cultivation over 40 days developed from an initial weight of 5–20.54 and 13.26 kg by using floating cage and longline, respectively (**Figure 4**).

The growth trials of *K. alvarezii* also showed different results for initial weights of 5 kg. In the process of cultivation for 40 days, the growth of this species increased to 19.2 and 12.7 for cultivation using floating cage and longline, respectively (**Figure 5**).

The results of the growth of new thallus with initial weight of 100 g showed the difference in the floating cage and on longline. From the results of the cultivation for 40 days, it was seen that the change in the weight of *E. denticulatum* on the longline was only around 432 g, while the change in weight in floating cage could reach an average of 1028 g. This result was very significant where changes in seaweed weights can reach 10 times for adequate seedlings (**Figure 6**).

There was also a difference between cultivations of *K. alvarezii* seeds in floating cage and on longline. With an initial 100 g wet weight, the change in wet

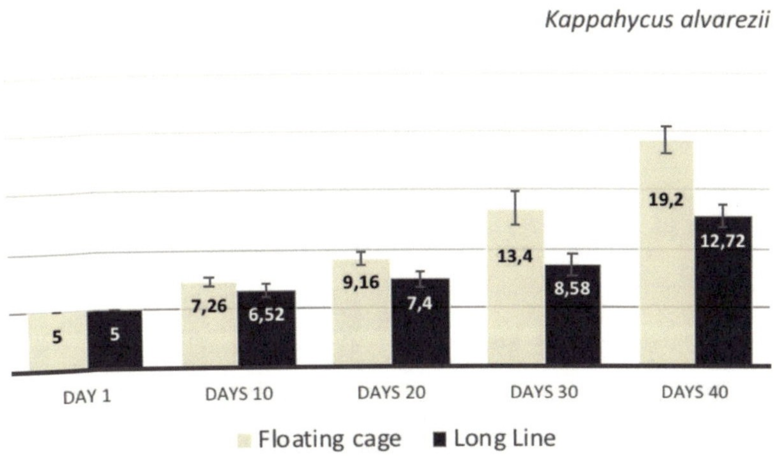

Figure 5.
Total growth of seaweed K. alvarezii *which was cultivated with floating cage and longline at an initial weight of 5 kg.*

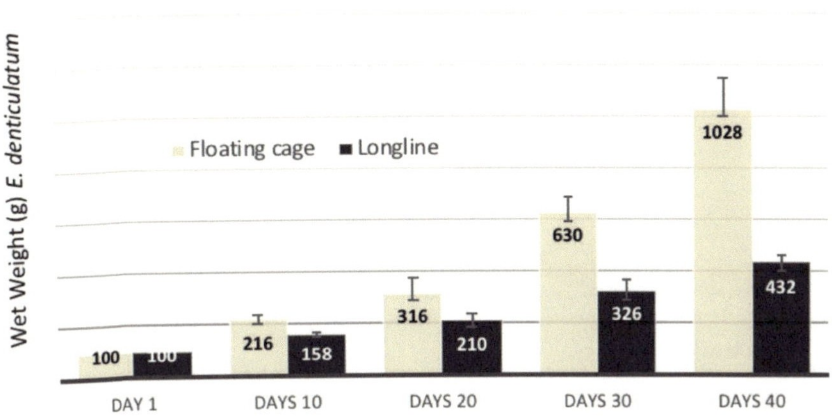

Figure 6.
Differences in growth of E. denticulatum *cultivated by floating cage and longline with initial weights of 100 g.*

weight of *K. alvarezii* by the longline method could reach 488 g, while with floating cage, it could reach an average of 758 g. This difference in growth was greatly influenced by several factors, including pests that were very intensely attacking seaweed, current pressure that could occasionally wash seaweed from its bonds, and seaweed *Chaetomorpha* sp. that developed quite high in all cultivation locations (**Figure** 7).

The difference in growth and weight of thallus in each month shows the difference between cultivation with floating cage and longline. On the growth chart, it can be seen that changes in seaweed weights had begun since the cultivation period for 10 days and the first 20 days. Furthermore, the changes were seen to be greater on the 30th and 40th days. This proves that maintenance with longline guarantees that seaweed can continue to grow well without any interference from pests and diseases that can reduce seaweed production. Another reason for the difference is that the longline cultivation method is prone to current shocks that can break the seaweed so that it will be washed away by the current.

Specific growth rates of *E. denticulatum* species differed in longline and floating cage cultivation methods. The average SGR of *E. denticulatum* was around 4.3%/day for the use of longline method, while for the use a floating cage, the SGR averaged

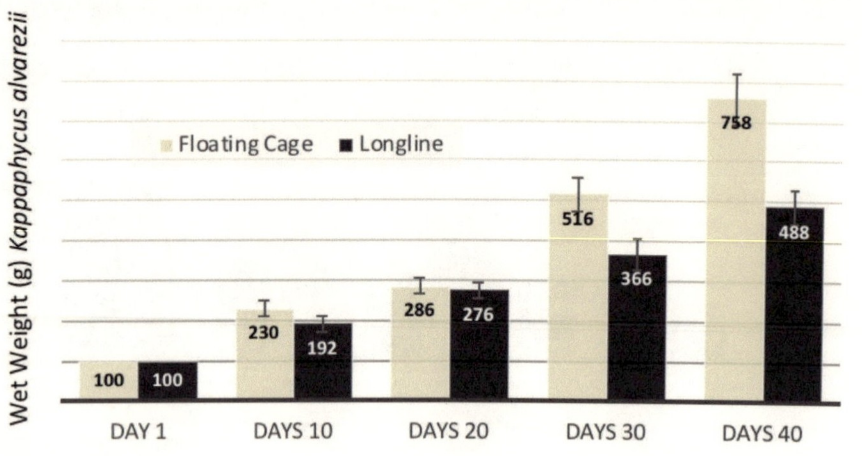

Figure 7.
Differences in growth of K. alvarezii *maintained by floating cage and longline with initial weights of 100 g.*

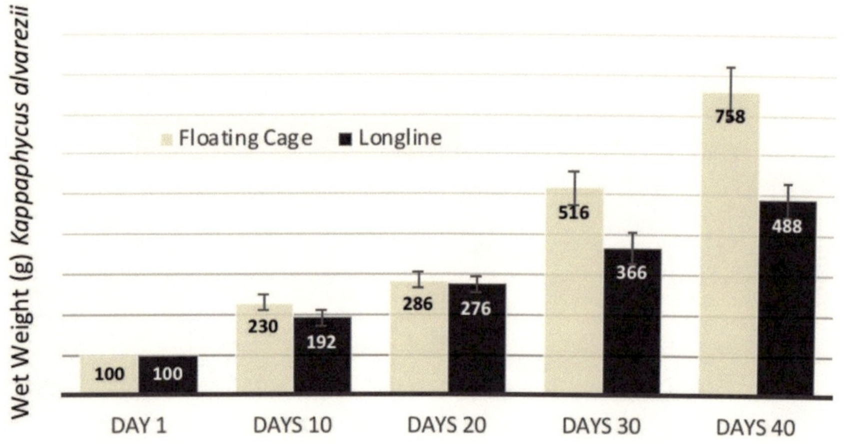

Figure 8.
Specific growth rate of E. denticulatum *cultivated with longline and floating cage.*

6.4%/day. SGR difference looked even each month. During the 6-month cultivation period (April–September) with a 35-day planting period, it was seen that the difference between floating cage and longline was around 3% (**Figure 8**).

The result of specific growth rates of seaweed *K. alvarezii* cultivated with longline was 4.1% per day, appearing to be lower on average than the specific growth rate of seaweed maintained with floating cage, which could reach 5.8%/day (**Figure 9**). The specific growth rate of *K. alvarezii* is an illustration of the daily growth rate of seaweed compared to the total growth in one harvest period. This

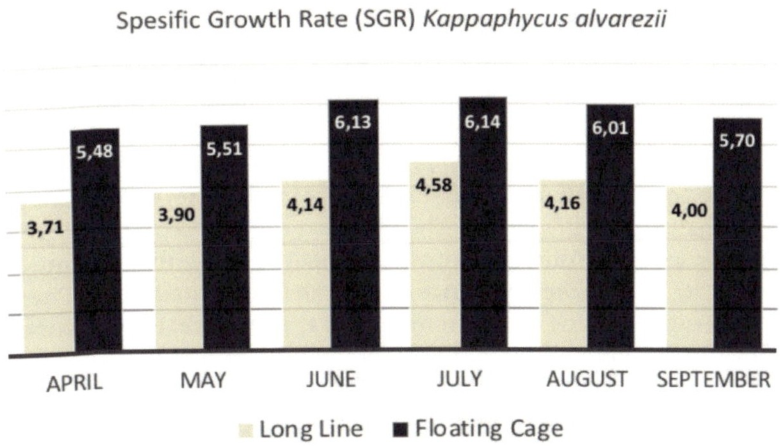

Figure 9.
Specific growth rate of K. alvarezii *that was cultivated by longline and floating cage.*

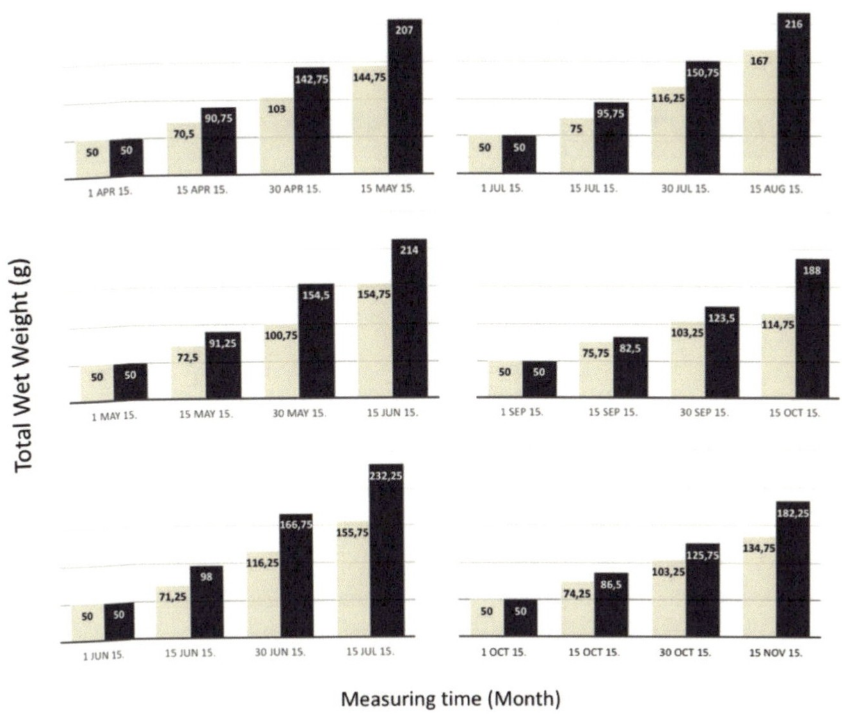

Figure 10.
Growth of thalli of E. denticulatum *cultivated with longline and floating net rafts [29, 30].*

means that the daily growth rate of seaweed maintained by longline is much lower than the daily growth rate of seaweed maintained by floating cages.

The growth of new thallus per month during the research from May to November showed that there were variations in growth each month that were quite different between the two methods. However, these variations did not provide a significant difference in thallus growth each month in the floating cage. The growth from May to July was seen to be different from the growth from September to November (**Figure 10**). The total growth using floating cage method was much higher than the growth using the longline method, especially in August to November. This can be expected from the presence of herbivorous fish that often attack seaweed in the months of August to November. The longline method is an open method that provides opportunities for herbivorous fish to eat seaweed, while floating cage is a closed method that protects seaweed from various attacks of pests, especially herbivorous fish.

The growth of the new thallus from May to November showed that there were variations in growth each month. However, these variations did not provide a significant difference for thallus growth each month in floating cage. Observation during May to July, growth was seen to be different from September to November (**Figure 11**).

However, if we look at the difference of growth in the floating cage method and the longline method for each different month, it appeared that the two methods had significantly different growths. The total growth in floating cage was much higher than the growth in the longline, especially in August to November. This can be expected from the presence of herbivorous fish that often attack seaweed in the months of August to November. It is important to be noted that longline method is an open method that provides opportunities for herbivorous fish to eat seaweed,

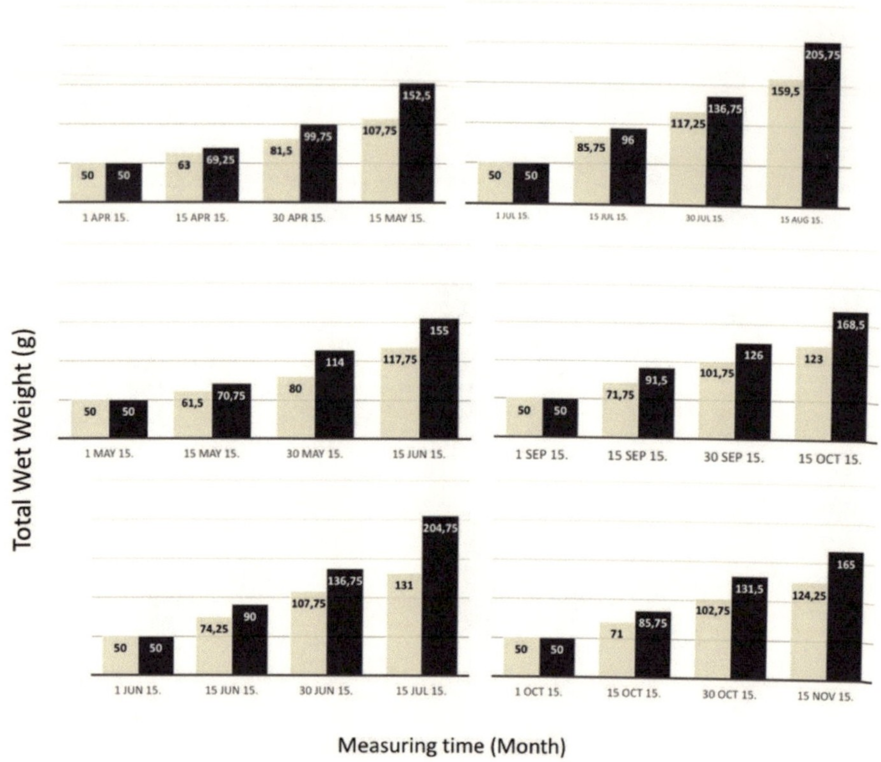

Figure 11.
Growth of the thalli of K. alvarezii *cultivated with longline and floating net rafts [29, 30].*

while the floating cage method is a closed method that protects seaweed from various attacks of pests, especially herbivorous fish (**Figures 10** and **11**).

The growth rate of *K. alvarezii* which was cultivated by floating cage was different from the growth of the same species in the longline. From the results of research conducted, the growth of *K. alvarezii* on longline was slower than the growth of the species in floating cage for 40 days. On the growth chart, the trend is seen to be rising and different between the two methods. The more visible difference was the growth in 40 days during the cultivation process. Production of *E. denticulatum* and *K. alvarezii* seemed to be slow with the longline method. Some of the indications seen at the cultivation site were that the use of the longline method in the harsh conditions of the waves experienced a hard shock and caused the fracture of the thallus of *K. alvarezii* and *E. denticulatum*. Broken thallus will reduce the weight achieved in 40 days of cultivation. The use of floating cage can avoid the drift of the thallus if it is broken and causes seaweed to remain in the cultivation area. Another situation was the extent that was still in the area of cultivation attacked by herbivorous fish. The most recent indication was that the average growth rates of the thallus of *K. alvarezii* and *E. denticulatum* in August were seen to decrease dramatically when fish herbivores were found to be abundant around the cultivation area. Production of *K. alvarezii* cultivated in Igang, Guimaras, Philippines, during February and March was 862 and 575 g/m/line. The growth rate of *K. alvarezii* was seen to increase during April and May with 1877 and 2237 g/m/line [26].

The increase in growth of this type is quite significant because *K. alvarezii* is kept in bamboo cages that block access of pest fish to attack *K. alvarezii*. The condition of seaweed in this area during the months of April and May looks very good. The time period of April to May is a good growing season. Some thalli are seen growing well and in good morphological form. While in Bongao, Southern Philippines, the growth rate of *K. alvarezii* can actually reach 300% of the average initial weight after being cultivated for 4–7 weeks [32], in Vietnam, the growth rate of *K. alvarezii* cultivated by the longline method with different depths, especially at a depth of 0.5–1 m, appeared to be increasing and in good condition during the time period of January–August, with a range daily growth of 6.14–6.26% on day 1. However, the growth rate increased in the time period of May–June in the range of 9.14–10.8% each day [33]. In Ubatuba Bay, Sao Paulo, Brazil, *K. alvarezii* was bred with the monoline method and grew well on the surface of the water and at a depth of 0.5 m [35]. The average growth of this species with a cultivation time of around 28 days was in the range of 5.2–7.2% day 1. However, growth was seen to decrease during cultivation for 59 days. Highest productivity was on day 44 during cultivation [34].

In Madagascar, the highest growth rate of *E. denticulatum* was recorded in April and February, while in March, decreased growth rate was caused by herbivorous fish and ice–ice disease [36]. Specific growth rate recorded at each planting season was 2.2% day 1 [32]. In Yucatan State, Mexico, the growth rate of *Eucheuma isiforme* increased during the 25-day cultivation of 2.21% day 1. The growth rate highly decreased in cultivation period after 50 days since the presence of pests and ice–ice disease [33, 37]. At Vizhinjam village, Kerala, India, the growth and production of *K. alvarezii* in shallow waters seemed high at 45–60 days of cultivation with the production of 24 and 36 kg, respectively, in March and May [34].

5. Thallus health condition in a floating cage

Thallus health can be seen from the morphological form of the thallus. The healthy shape of the thallus morphology looks cleaner. Healthy thallus has tapered tip, and there are many new thalli starting to emerge. Thallus of seaweed

Figure 12.
Thallus of K. alvarezii *cultivated by floating cage.*

maintained by the floating cage looked cleaner than that maintained by longline (**Figure 12**). At the time of a fairly high pest attack, seaweed thallus looked white and would eventually die. Seaweed thallus which was attacked by pests had many wounds and bleached and the thallus was cut off. The intensity of this pest attack was seen at all times, especially during the attack of *Siganus* sp., which usually happens in August–November every year. However, this pest is not the only one that often attacks seaweed. Some farmers reported that their seaweed was attacked by sea turtles. This is because sea turtle is a herbivore that likes to eat seaweed.

Besides fish pests, it was also reported that a number of thalli that died or were sick were mainly caused by ice–ice disease which often attacks seaweed. In general, longline cultivation provides opportunities for seaweed to be affected by changes in temperature and salinity at certain times, especially in the rainy and dry seasons. During summer, a lot of seaweed is affected by surface heat so that it becomes stressed and eventually bleaches. When the rainy season occurs, some seaweed will be affected by changes in surface salinity, which drops suddenly due to rainwater, because seaweed is very close to the surface of the water. Different things can be seen with the use of the floating cage. Seaweed in this tool looks fresher and healthier because it will be able to avoid herbivorous pests. Besides that, seaweed will also be able to avoid changes in temperature and surface salinity during rainy season or dry season. Seaweed in floating cage will be submerged at depths above 10 cm below sea level. Thallus of seaweed maintained by floating cage looks clean, and the tip of the thallus is tapered, and there are no bites of data pests nor is there ice–ice disease [38]. Seaweed in floating cages has many thalli that continue to grow. Thallus morphologies of *E. denticulatum* and *K. alvarezii* that were cultured with longline and floating cage looked different in August and September. In August, the intensity of herbivore attack was seen to be high enough so *E. denticulatum* and *K. alvarezii* were highly consumed by herbivorous fish (**Figure 1**). In September, it could be seen that epiphytic *Chaetomorpha* sp. wrapped *E. denticulatum* and *K. alvarezii* until they could not be seen. Morphologies of *E. denticulatum* and *K. alvarezii* that were in floating cage looked intact, and there was no sign of being consumed by fish. Growth of *Chaetomorpha* sp. on each thallus of *K. alvarezii* (weight

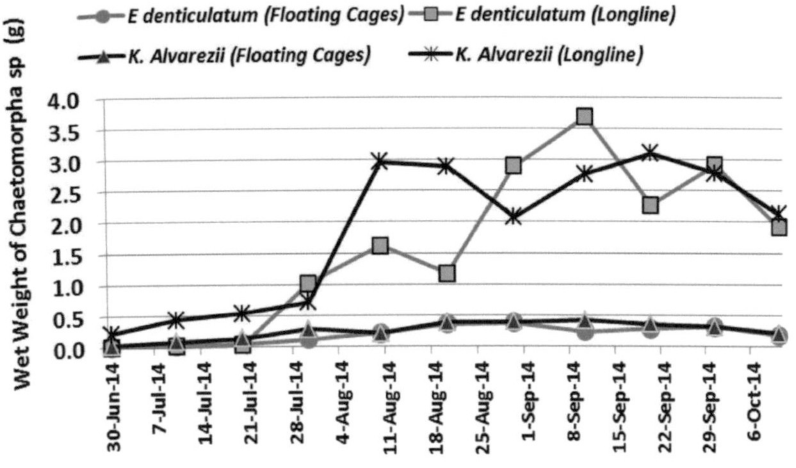

Figure 13.
Growth of Chaetomorpha *sp. during cultivation periods on longline and inside the floating cages.*

100 g) that was cultured by longline was seen to rise at the beginning of August and at the end of September with 2.9, and 2.78 g, respectively. Meanwhile, growth of *Chaetomorpha* sp., which covered *E. denticulatum* also at longline, was seen to be very high at the end of August and at beginning of September with 3.68 and 2.89 g, respectively. Different phenomena were seen in floating cage where *Chaetomorpha* sp. growth did not exceed an average of 0.5 g (**Figure 13**).

6. Floating cage development recommendations

The results showed that floating cage had a significant effect on increasing seaweed production. Seaweed is completely protected from pests and ice–ice disease. Increased seaweed production will provide increased harvest income from seaweed cultivation. Floating cage is a very simple innovation tool that has a series of seaweed cultivation tools from PVC paralon and nets. This tool is very environmentally friendly if compared with the current seaweed method. At present, the longline method has various serious problems. The problem is, in certain areas, pests and epiphyte attachments occur as well as the use of used plastic bottles. The use of used plastic bottles is an act that is not environmentally friendly. Used plastic bottles used as buoys are seen to pollute the ocean and have a wider impact on organisms around the cultivation area (**Figure 14**).

The survey results explain that in 1 Ha longline cultivation area will require at least 200 used plastic bottles. Plastic bottles are used as buoys in seaweed cultivation. Used plastic bottles are taken from bottle wastes in the community environment. The use of used plastic bottles can be used only in 1 year of use; after 1 year usually these plastic bottles will leak and be thrown into the sea. Many serious problems occur among farmers. At present it is very important to develop an alternative seaweed cultivation that is environmentally friendly and has high productivity.

Floating cage that uses multifilament nets and PVC pipe does not cause environmental damage (**Figure 14**). PVC pipe is used as a float and also as a main pole to hold the net. PVC pipe series will last long enough in the waters. The use of PVC pipe and nets with good care will provide a long usage period. Arrangement of floating cage in the sea will look neatly arranged. The prospect of using floating cage has a very good future with various benefits. We got the advantage of using floating cage by interviewing farmers who use this tool in several seaweed

Figure 14.
Longline cultivation using used plastic bottles (A) and clean and beautiful floating cage arrangement (B).

cultivation locations in Southeast Sulawesi, Indonesia. The results of our interviews are formulated in order of priority of the benefits obtained by the community in using floating cage tools. These advantages include:

- *Easy to assemble*. This tool is made of a PVC pipe with a size of 3 mm which can be designed rectangular with an appropriate connection. PVC pipes are cut according to the desired size and connected with the appropriate connections (**Figure 1**).

- *Durable*. This tool is proven to be resistant at sea and will never rust. The durability of the tool with good care can reach 20–25 years of use.

- *Easy to maintain*. This tool has a very simple form and will be easily cleaned and maintained to extend its durability. Optimal results can be provided, it takes one reserve outer net as a replacement net each time the cultivation.

- *Float and balanced*. The shape of this tool is simple and floats on the surface of the water. This tool can be placed in all marine topography.

- *Full protection*. This tool is effective in protecting seaweed. The outer net as a protective device is very effective in avoiding herbivorous attacks that often graze seaweed.

- *Clean seaweed*. Seaweed will be cleaned from various dirt that are attached to the sea, including epiphyte, with a note that farmers must also often do the cleaning on the outer wall and in the net.

- *Volume of large seaweed*. The number of cultivated seaweed seeds is large with an initial weight capacity of 40–80 kg wet. After 30 days, it can reach 160–320 kg of wet confinement.

- The location of cultivation is narrow and orderly. The design of this tool does not require a large cultivation location. With just a small and regular area, the farmer can arrange the floating cage placement.

The many advantages of using floating cage tools illustrated very clearly that the future use of floating cage methods will be very good. Although currently the use of floating cage is still limited to certain locations, however, this development will continue to occur along with the desire to use prospective tools with excellent economic and ecological benefits. Cultivation communities will be able to choose to use tools that are economically and ecologically beneficial. The ecological advantage is in favor of the sustainability of other marine resources.

7. Conclusion

Innovation of cultivation methods using floating cage gives good results to support the growth of *Kappaphycus alvarezii* and *Eucheuma denticulatum*. The growth of *K. alvarezii* looks quite good with a specific growth rate up to 6.4%/day using floating cage method. However, *K. alvarezii* only has 4.2% growth/day using the longline method. The floating cage method can truly protect seaweed from herbivorous fish. Thus, the growth of seaweeds *K. alvarezii* and *E. denticulatum* is better because they are not attacked by fish.

Acknowledgements

I am grateful to the Ministry of Research, Technology and Higher Education of the Republic Indonesia, for the full research funding of this research project and to the Faculty of Fishery and Marine Science, Halu Oleo University, for the sample analysis and literature assistance.

Author details

Ma'ruf Kasim[1*], Abdul Muis Balubi[1], Ahmad Mustafa[1], Rahman Nurdin[1], Rahmad Sofyan Patadjai[1] and Wardha Jalil[2]

1 Faculty of Fishery and Marine Science, Halu Oleo University, Kendari, Southeast Sulawesi, Indonesia

2 Faculty of Fishery, Dayanu Iksanuddin University, Baubau, Southeast Sulawesi, Indonesia

*Address all correspondence to: marufkasim@uho.ac.id

References

[1] Azanza-Corrales R, Mamauag SS, Alfiler E, Orolfo MJ. Reproduction in *Eucheuma denticulatum* (Burman) Collins and *Kappaphycus alvarezii* (Doty) Doty, in Danajon reef Philippines. Aquaculture. 1992;**103**:29-34

[2] Sulu R, Kumar L, Hay C, Pickering T. Kappaphycus seaweed in the Pacifik: Review of Introduction and Field Testing Proposed Quarantine Protocols. The Institute of Marine Resources, The University of the South Pacific; 2003. 84 p

[3] Doty MS, Alvarez VB. Seaweeds farm: A new approach for U.S. industry. In: Proceeding of the 9th Annual Conference Proceeding. University Hawaii; 1973. pp. 701-707

[4] Doty MS. Farming the red seaweed, Eucheuma for carrageenans. Micronesica. 1973;**IX**(1):59-73

[5] Parker HS. The culture of the red algal genus Eucheuma in the Philippines. Aquaculture. 1974;**3**:425-439

[6] Dawes CJ. Irradiance acclimation of the cultured Philippine seaweeds, *Kappaphycus alvarezii* and *Eucheuma denticulatum*. Botanica Marina. 1992;**35**:189-195

[7] Azanza-Corales R, Aliaza TT, Montano NE. Recruitment of Eucheuma and Kappaphycus on a farm in Tawi-Tawi, Philippines. Hydrobiologia. 1996;**326/327**:235-244

[8] Kasim M, Asnani A. Penentuan musim reproduksi generative dan preferensi perekatan spora rumput laut Eucheuma cottoni. Ilmu Kelautan. 2012;**17**(4):209-216

[9] Prakash J. Fiji. In: Adams T, Foscarini R, editors. Proceeding of Regional Workshop on Seaweeds Culture and Marketing. South Pacific Aquaculture Development Project; 14-17 November 1989, Suva, Fiji. Food and Agriculture Organisation of the United Nations; 1990. pp. 1-9

[10] Luxton DM, Robertson M, Kindley MJ. Farming of Eucheuma in the South Pacific islands of Fiji. Hydrobiologia. 1987;**151/152**:359-362

[11] Gentle T. Tuvalu. In: Adams T, Foscarini R, editors. Proceeding of Regional Workshop on Seaweeds Culture and Marketing. South Pacific Aquaculture Development Project; 14-17 November 1989, Suva, Fiji. Food and Agriculture Organisation of the United Nations; 1990. pp. 32-33

[12] Adnan H, Porse H. Culture of *Eucheuma cotonii* and *Eucheuma spinosum* in Indonesia. Hydrobiologia. 1987;**151-152**:355-358

[13] Luxton DM. Aspect of farming and processing of Kappaphycus and Eucheuma in Indonesia. Hydrobiologia. 1993;**260/261**:365-371

[14] Wu CY, Li JJ, Xia EZ, Peng ZS, Tan SZ, Li J, et al. Transplant and artifisial cultivation of *Eucheuma striatum* in China. Oceanol. Limnol. Sin. 1988;**19**:410-417

[15] Tanaka H. Foreword. In: Adams T, Foscarini R, editors. Proceeding of Regional Workshop on Seaweeds Culture and Marketing. South Pacific Aquaculture Development Project; 14-17 November 1989, Suva, Fiji. Food and Agriculture Organisation of The United Nations; 1990. pp. iii-iiv

[16] Smit AJ, Fourie AM, Robertson BL, Du Preez DR. Control of the herbivorous isopod, *Paridotea reticulata*, in *Gracilaria gracilis* tank

cultures. Aquaculture. 2003;**217**(1-4): 385-393. DOI: 10.1016/S0044-8486 (02)00412-X

[17] Lirasan T, Twide P. Farming Eucheuma in Zanzibar, Tanzania. Hydrobiologia. 1993;**260-261**:353-355

[18] Ohno M, Nang HQ, Hirase S. Cultivation and carrageenan yield in quality of *Kappaphycus alvarezii* in the water of Vietman. Journal of Applied Phycology. 1996;**8**:431-437

[19] de Paula EJ, Pereira RTL, Ohno M. Strain selection in *Kappaphycus alvarezii* var. alvarezii (Doty) Doty ex P. Silva (Rhodophyta, Solierieaceae) using tetraspore progeny. Journal of Applied Phycology. 1999;**11**(11):111-121

[20] Rincones RE, Rubio JN. Introduction and commercial cultivation of the red alga Eucheuma in Venezuela for the production of phycocolloids. World Aquaculture Magazine. 1999;**30**(2):57-61

[21] Munoz J, Freile-Pelegrin Y, Robledo D. Mariculture of *Kappaphycus alvarezii* (Rhodophyta Solieriaceae) color strain in tropical water of Yucatan, Mexico. Aqua. 2004;**239**:161-177

[22] Kumar KS, Ganesan K, Rao PVS, Thakur MC. Seasonal studies on field cultivation of *Kappaphycus alvarezii* (Doty) Doty on the northwest coast India. Journal of Applied Phycology. 2015. DOI: 10.1007/s 10811-015-0629y

[23] Goromel de Goes H, Reis RP. An initial comparison of tubular netting versus tie-tie methods of cultivation for *Kappaphycus alvarezii* (Rhodophyta, Solieriaceae) on the south coast of Rio de Janeiro state, Brazil. Journal of Applied Phycology. 2011;**23**:607-613

[24] McHugh DJ. Prospects for seaweed production in developing countries. In: FAO Fisheries Technical Circular. No. 968. Rome: FAO; 2002. pp. 28-29

[25] Tomas F, Cebrian E, Ballesteros E. Differential herbivory of invasive algae by native fish in the Mediterranean sea. Estuarine, Coastal and Shelf Science. 2011;**92**:27-34. DOI: 10.1016/j.ecss.2010.12.004

[26] Ganesan M, Thiruppathi S, Sahu N, Rengarajan N, Veeragurunathan V, Jha B. In situ observations on preferential grazing of seaweeds by some herbivores. Current Science. 2006;**91**:1256-1260

[27] Hurtado-Ponce AQ. Cage culture of *Kappaphycus alvarezii* var. tambalang (Gigartinales, Rhodophyceae). Journal of Applied Phycology. 1992;**4**(4): 311-313. DOI: 10.1007/BF02185787

[28] Biswajit D, Suresh KM, Syda RG. Integration of Seaweed (*Kappaphycus alvarezii*) and Pearl Oyster (*Pinctada fucata*) along with Asian Seabass (*Lates calcarifer*) in Open Sea Floating Cage off Andhra Pradesh Coast. National Fisheries Development Board; 2009. http://eprints.cmfri.org.in/6070/1/10._Bis.pdf (diakses 12 Mei 2019)

[29] Kasim M, Mustafa A, Munier T. The growth rate of seaweed (*Eucheuma denticulatum*) cultivated in longline and floating cage. AACL Bioflux. 2016;**9**(2):291-299

[30] Kasim M, Mustafa A. Comparison growth of *Kappaphycus alvarezii* (Rhodophyta, Solieriaceae) cultivation in floating cage and longline in Indonesia. Aquaculture Reports. 2017;**6**:49-55. DOI: 10.1016/j.aqrep.2017.03.004

[31] Kasim M, Jamil MR, Irawati N. Occurrence of macro-epiphyte on *Eucheuma spinosum* cultivated on floating cages. AACL Bioflux. 2017;**10**(3):633-639

[32] Kasim M, Asjan B, Effendy IJ, Wanurgayah IE. Influence of initial

weight of seeds in variation of growth and carrageenan content of *Eucheuma spinosum*. AACL Bioflux. 2018;**11**(4):1155-1163

[33] Villanueva RD, Romero JB, Montano MNE, de la Pena PO. Harvest optimization of four *Kappaphycus* species from the Philippines. Biomass and Bioenergy. 2011;**35**:1311-1316

[34] Ohno M, Largo DB, Ikomoto T. Growth rate, carrageenan yield and gel propery of culture kappa-carrageenan producing red alga *Kappaphycus alvarezii* (Doty) Doty in the subtropical water oh Shikoku, Japan. Journal of Applied Phycology. 1994;**1994**(6):1-5

[35] Hayashi L, Oliveira EC, Bleicher-Lhonneur G, Boulenguenguer P, Pereira RTL, von Seckendorff R, et al. The effect of selected cultivation condition on the carrageenan characteristics of *Kappaphycus alvarezii* (Rhodophyta, Solieriaceae) in Ubatuba Bay, Sao Paulo, Brazil. Journal of Applied Phycology. 2007;**19**:505-511

[36] Mollion J' Braud JP. A Eucheuma (Solieriaceae, rhodophyta) cultivation test on the south-west coast Madagascar, Hydrobologia 1993;260/261:373-378

[37] Perez-Enriquez R. Growth of *Eucheuma isifonia* (*C. Agardh*) J. Agardh on experimental raft off the coast of Yucatan state, Mexico. Journal of Applied Phycology. 1996;**8**:27-28

[38] Budiyanto, Kasim M, Abadi SY. Growth and carrageenan content of local and tissue culture seed of *Kappaphycus alvarezii* cultivated in floating cage. AACL Bioflux. 2019;**12**(1):167-178

Chapter 9

Insects in Aquaculture Nutrition: An Emerging Eco-Friendly Approach or Commercial Reality?

André Freccia, Jiovani Sergio Bee Tubin, Artur Nishioka Rombenso and Maurício Gustavo Coelho Emerenciano

Abstract

Animal-origin food production presents an accelerated growth worldwide due to an increase in human demand. The aquaculture sector is one of the major players in terms of volume of animal protein production, and the availability of feedstuff to supply aquaculture feed (aquafeed) chain will be one of the main challenges for the next decades. Aquafeeds are mostly based on cereals, oilseeds, and marine-origin ingredients. The competition for feedstuff from the terrestrial animal industries such as pet, poultry, and swine challenges the profitability of aquafeeds, and complimentary ingredients need to be found. Many studies have focused on alternative protein sources, but the benefits of plant proteins, microorganisms-based, and diverse animal by-products are still under intense investigation to address some constraints including antinutritional factors and unbalanced nutrient profile. In this sense, the use of insects on the nutrition of aquatic animals could be an alternative. This chapter was elaborated to be an introductory reading for both academic and private sector and will discuss (i) the benefits of insects in animal nutrition, (ii) elucidate the nutritional aspects of different insect meals, (iii) bring some practical developments on aquatic nutrition, and finally (iv) discourse about constraints on insect use and its future perspectives.

Keywords: animal nutrition, alternative feedstuff, additives, lipid, protein, chitin

1. Introduction

The demand for meat in 2050 will be 58% higher than the demand in 2010 [1], and revolutions in the form of animal protein production should happen to supply this demand; this will mean greater pressure on food resources especially on the ingredients for the formulation of rations [2]. In the quest for more sustainability in long-term animal production, the search for alternative ingredients is essential, since conventional ingredients such as soybean meal, wheat, corn, and animal meal have large price swings and exponentially high values years after years [2].

In this context, insect breeding is an alternative as a source of nutrients for animal feed and is a way to increase food and feed safety as an important source of protein [3]. In addition, insect breeding is considered a sustainable production, since these small animals feed on agroindustrial tailings, various organic wastes [3], and even inorganic ones [4].

The production of food of animal origin presents an accelerated growth in the world specially the chicken, swine, and cattle production [1]. These are produced with cereal and oilseed diets [2]. The nutrition of these animals requires high protein and energy levels that could be supplied by the use of insects. In addition, more than 1 million species of insects have been cataloged, and it is estimated that there are between 5 and 10 million, that is, only 10–20%, are known [5]; more than 1500 are used as food for about 3000 ethnic groups in more than 120 countries [6].

This scenario results in an emerging need for further studies to identify potential species to be produced and used in animal nutrition and as addressed by the FAO [7] as an alternative to traditional sources for human consumption. With these perspectives, it is exposed to the needs of recycling our ideas of human and animal feeding. In animal production, the mystification of this practice occurs more quickly, since many of the production animals, such as birds and fish, have the habit of feeding on insects.

2. Insects in animal nutrition

Agricultural production in the world faces many challenges to meet the growing demand for animal products, such as the demand for meat, which in 2050 will be 58% higher than the demand in 2010 [1]. Many studies have focused on new nutritional perspectives, but the benefits and weaknesses of vegetable protein sources and animal by-products are still under intense investigation. Ingredients of plant origin, for example, have several adverse effects on animal performance, attributable, among others, to antinutritional factors, inadequate profile of fatty acids, and amino acids [8, 9]. However, both are associated with environmental problems (e.g., exploitation of natural resources), economic (fluctuation of feed prices), or production (variation of quality and quantity) [10].

When considering the long-term sustainability of animal production, evaluations of alternative ingredients are essential because conventional ingredients, such as soybean meal and fishmeal [2], are raw materials that have been unstable due to the demand of other segments such as birds, swine, and pet [11]. Along with their derivatives and by-products, animal meal and soybean meal are two of the most widely used protein sources as feed ingredients. However, both are associated with environmental (e.g., exploitation of natural resources), economic (fluctuations in feed prices), or production (variation of quality and quantity) problems [10]. In this context, many studies have addressed new nutritional perspectives with alternative ingredients. However, the benefits and weaknesses of these sources are still the subject of intense research [11].

Supported by the problems mentioned above, it is necessary to search for alternatives. In this context, the use of insects could be a source of protein for animal feed. The commercial scale production of insects could guarantee a constant production in quantity, quality, and price. The use of this source to feed terrestrial animals and aquatic organisms already has some premises [7, 12] and brings, among others, the following advantages: (i) nutritional quality, (ii) noncompetition with food resources or land use, and (iii) use of "nutrients from residues" or "energy leftovers" from agro-food production as nutritional sources for insect growth. These small organisms can be grown in industrial plants and do not need large areas, especially when compared to other crops such as soybeans.

The use of insects contributes to the natural recycling of nutrients and could be a source of high-quality animal protein derived from environmentally sustainable technology. In addition, a survey conducted in northern Italy showed that 90% of consumers would buy fish fed with insect meal [13]. This makes perfect chain fitting ("agriculture + fish + consumer") in addition to a more sustainable production call. In this item it is worth mentioning that some insects have the capacity to reduce the final biomass of residues in 50%, and specifically of nitrogen residues in 30–50%, and phosphorus in 61–70% [3, 14]. In other words, they convert organic residues into high-quality fertilizers in addition to forming "protein biomass" with admirable figures (~40% crude protein and ~ 30% lipid) [15, 16]. The arguments that reinforce its use would be that insects grow and reproduce easily, have high feed conversion efficiency, and can feed on organic wastes [17].

In addition, studies have shown that it is technically feasible to produce large-scale insects and use them as a sustainable protein alternative in the diet of birds, swine, cattle, and aquatic animals [12]. Once mass production of these small animals has been achieved, it would be possible to lower their cost and achieve economic viability in replacing traditional protein ingredients in animal feed [2]. Today the cost per kilo of the flour of some species of insects in the Brazilian market reaches more than US$50.00/kg. In the literature we have some studies that used insects to feed fish such as house fly larvae (*Musca domestica*) as a source of protein for feeding tilapia and African catfish (*Clarias gariepinus*) [18–20]; larvae of butterflies (*Bematistes macaria*) for feeding African catfish hybrids (*Heteroclarias*) [21]; and *Tenebrio molitor* in the diet of African catfish [22], goldfish (*Sparus aurata*) [23], rainbow trout (*Oncorhynchus mykiss*) [24], European sea bass (*Dicentrarchus labrax*) [25], Nile tilapia (*Oreochromis niloticus*) [10], and cheap meal (*Nauphoeta cinerea*) [26].

The larvae and pupae of the *Tenebrio molitor* beetle are promising options with several studies mainly for feeding fish and birds. This short-lived, easy-to-breed insect would be an alternative to temperate Western countries where this small animal is endemic [27]. Information on the breeding, feeding, and nutrient needs of this insect is already available in the literature [28, 29]. The larvae of *T. molitor* are omnivorous but are usually fed wheat flour or meal and supplemented with soybean meal, skimmed milk powder, or yeast [1]. The moisture in the feed seems to be fundamental for *T. molitor* because it can affect the productivity and fat content [30]. However, breeding of *T. molitor* larvae with resources such as wheat flour, soybean, and skimmed milk cannot be considered sustainable, since these products could be considered more suitable for direct consumption for humans or used in the nutrition of domestic animals.

The larvae of *T. molitor* have the ability to recycle vegetable waste of low nutritional quality and turn them into high-quality food (biomass) [1]. Tea bag larvae were fed on food leftovers and turned this waste into a high-quality protein source, which reiterates the potential of tenacious larvae as a promising and sustainable alternative protein source [31]. In addition, it has been recently reported that this animal can even recycle plastics (**Figure 1**) because it has specific bacteria in its tract capable of degrading this material [4]. Thus, it would be possible to solve two major problems of the contemporary world: the scarcity of resources for human and animal food, besides the "biorecycle" of plastics. Another insect that has stood out with high potential for use in animal nutrition is the cinerea cockroach (*Nauphoeta cinerea*). Some companies in South America already produce it on a commercial scale to attend to zoos and feed rations mainly for birds. Due to its high protein content (approximately 60%), the demand for this flour has increased over the years. Some studies have demonstrated the potential of its use in fish diets [26].

Figure 1.
Adult and larvae of Tenebrio molitor *(A) and feeding with plastic (B). Source: A, available at <plantascarnivoras.com.br> access 09/20/2016; B - [4].*

3. Nutritional aspects of insect meals

The proper way to go for the use of alternative ingredients in animal feed is to understand their nutritional characteristics (physical, chemical, and biological). Knowledge of these traits will be of paramount importance in assessing the potential for their use in animal feed, either as a substitute or complimentary. It is necessary to evaluate these ingredients in the diet of the animals, considering digestibility, performance, nutrient balance, carcass characteristics, economic results, and sustainability in the production chain.

The potential of insect use in animal and human food is mainly because there are more than 1 million known species. This generates innumerable possibilities and alternatives for its use; however, many studies are necessary since after identifying a species with potential, strategies should be created for production, reproduction, genetic evolution, and processing. The nutritional composition of several insect meals was compiled and divided into amino acid, bromatological composition, and fatty acid profile (**Table 1**).

Insect meal may have a high content of ethereal extract, and its variation influences the crude energy (kcal kg^{-1}) of the diets, the energy:protein ratio, as well as the ethereal extract of the carcasses. Other authors have also reported high values of ethereal extract in insect meal that prevented the high inclusion of these in the diets [18]. High values of ethereal extract are premises for the occurrence of oxidation (rancidity) of fats, reducing the shelf life of this product. The inclusion of antioxidant additives in insect meals is suggested. A short-term solution would be to further develop preprocessing and manufacturing procedures to extract the excess lipid in the meals and then utilize it as a lipid source in feeds or in any other industry, approach already adopted to manufacture terrestrial animal by-product meals.

Some studies show large crude protein variation and ethereal extract between the same species, found 40% of crude protein and 25% of ethereal extract for tenebrio, lower value than other studies [24, 25] that reported crude protein values higher than 50%, but lower ethereal extract levels than in the present study. In this context, insects can be used as a source of protein and energy. The larval stage of insects usually has higher ethereal extract values, as these accumulate energy for metamorphosis. Its fatty acid profile is very variable, suggesting that as feed occurs modulation of the fatty acid profile of the insects, which may be a prelude to the inclusion of EPA (eicosapentaenoic acid, 20:5n-3) and DHA (docosahexaenoic acid, 22:6n-3) of lower quality ingredients. Roasted meal is an excellent source of protein (~66.84%), being superior to

Species	Stage	Proximate analysis (% dry matter)				Amino acid (% total)			Fatty acid (% total)	
		ASH %	EE %	CP %	NFE %	LYS	MET	THR	Satura.	Polyuns.
Phyllognathus excavates	Adult	7.8 ± 0.2	15.9 ± 1.4	65.7 ± 1.3	10.6 ± 0.1	6.34	1.42	4.1	28.7 ± 3.0	11.8 ± 0.4
Rhynchophorus ferrugineus	Larvae	6.6 ± 0.6	11.8 ± 1.5	34.6 ± 0.3	47.0 ± 1.3	6.18	0.45	4	42.5 ± 0.4	13.0 ± 0.4
Tenebrio molitor	Larvae	3.5 ± 0.2	30.1 ± 0.7	58.4 ± 0.4	8.0 ± 0.2	6.03	0.64	4.49	22.2 ± 0.1	31.5 ± 0.1
Zophoba morio	Larvae	2.5 ± 0.3	38.0 ± 0.3	53.5 ± 0.4	6.0 ± 1.1	5.82	0.76	4.33	38.8 ± 0.2	24.0 ± 0.0
Calliphora vicina	Larvae	8.0 ± 0.1	20.1 ± 0.7	48.3 ± 0.9	23.6 ± 0.1	7.99	2.16	4.86	28.5 ± 0.2	28.0 ± 0.1
Chrysomya megacephala	Larvae (L3)	7.2 ± 0.1	27.0 ± 3.2	61.8 ± 0.3	4.0 ± 3.4	8.53	2.22	4.51	35.9 ± 1.2	31.3 ± 0.7
Chrysomya megacephala	Pupae	6.1 ± 0.1	16.5 ± 0.0	46.8 ± 1.1	30.6 ± 1.1	7.87	2.76	5.02	35.4 ± 0.6	26.2 ± 0.1
Eristalis tenax	Larvae (L3)	13.9 ± 0.4	5.8 ± 0.6	40.9 ± 0.9	39.4 ± 1.1	8.45	2.37	5.02	41.7 ± 1.4	1.6 ± 0.0
Hermetia illucens	Larvae (L5)	9.3 ± 0.3	18.0 ± 1.6	36.2 ± 0.3	36.5 ± 1.0	7.6	1.5	5.39	67.1 ± 0.6	15.9 ± 0.6
Hermetia illucens	Pupae	19.7 ± 0.1	15.6 ± 0.1	40.7 ± 0.4	24.0 ± 0.7	7.31	3.26	4.95	65.8 ± 0.1	1.1 ± 0.0
Lucilia sericata	Larvae (L3)	4.9 ± 0.9	28.4 ± 1.5	53.5 ± 4.4	13.2 ± 4.6	7.66	3.36	5.38	27.8 ± 0.1	9.5 ± 0.4
Lucilia sericata	Pupae	4.9 ± 0.2	26.6 ± 1.0	59.0 ± 1.5	9.5 ± 0.1	7.91	3.08	4.6	28.8 ± 0.4	11.0 ± 0.1
Musca domestica	Larvae (L3)	6.5 ± 1.5	31.3 ± 1.6	46.9 ± 4.1	15.3 ± 4.0	8.36	3	4.87	32.6 ± 0.1	7.6 ± 0.1
Musca domestica	Pupae	8.4 ± 2.9	33.7 ± 0.7	40.1 ± 0.4	17.8 ± 0.3	7.57	3.44	5.28	30.0 ± 1.1	7.5 ± 0.4
Protophormia terraenovae	Larvae (L3)	3.9 ± 0.1	28.3 ± 0.6	46.3 ± 0.6	21.5 ± 0.1	8.23	2.3	4.78	27.1 ± 0.2	21.9 ± 0.2
Protophormia terraenovae	Pupae	8.8 ± 0.1	23.6 ± 0.3	56.0 ± 2.0	11.6 ± 2.2	7.89	2.55	4.83	26.6 ± 1.6	21.7 ± 0.2
Acheta domestica	Adult	5.6 ± 0.0	15.9 ± 0.2	73.1 ± 3.3	5.4 ± 0.3	6.16	1.49	4.1	34.2 ± 0.1	43.2 ± 0.1
Anacridium aegyptium	Adult	3.7 ± 0.1	17.6 ± 0.2	66.0 ± 5.0	12.7 ± 4.8	5.73	2.36	4.49	30.3 ± 1.7	30.0 ± 0.7
Gryllus assimilis	Adult	4.8 ± 0.1	23.2 ± 0.6	64.9 ± 0.5	7.0 ± 0.3	6.46	1.1	4.11	34.0 ± 0.6	37.5 ± 0.3
Heteracris littoralis	Adult	5.1 ± 0.1	8.8 ± 0.0	74.4 ± 1.0	11.7 ± 1.0	6.01	1.02	3.9	27.7 ± 0.6	42.1 ± 0.1
Locusta migratoria	Adult	4.0 ± 0.0	29.9 ± 0.5	58.5 ± 0.5	7.6 ± 0.1	6.33	0.54	4.28	36.4 ± 0.1	15.9 ± 0.4

Species	Stage	Proximate analysis (% dry matter)				Amino acid (% total)			Fatty acid (% total)	
		ASH %	EE %	CP %	NFE %	LYS	MET	THR	Satura.	Polyuns.
Fish meal	—	18.0 ± 0.2	8.2 ± 0.0	73.0 ± 0.8	0.8 ± 0.7	8.78	2.93	6.26	36.1 ± 1.1	37.3 ± 0.0
Soybean meal	—	7.8 ± 0.0	3.0 ± 0.0	50.4 ± 0.2	38.8 ± 0.3	6.34	1.01	4.17	24.0 ± 1.9	55.4 ± 0.8

EE, ethereal extract; CP, crude protein; NFE, nitrogen-free extract, includes fiber; LYS, lysine; MET, methionine; THR, threonine; Satura., saturated fatty acids, all fatty acids without double bonds; Polyuns., polyunsaturated fatty acids, all fatty acids ≥2 double bonds.

Table 1.
Proximate analysis, amino acids (lysine, methionine, and threonine), and fatty acid (saturated and monounsaturated fatty acids) of selected insects, fish meal, and soybean meal. Values were based on [33].

the main protein ingredients used in the formulation of diets for fish such as soybean meal, fishmeal, meat and bone meal, meal of viscera [32], and lower feather meal and blood meal, which rely on processing to improve their digestibility [32].

In terms of nutritional profile, the use of organic residues for insect production should be further investigated, especially when using foods with high levels of mycotoxins (residues). The mycotoxins, when is consumed by insects, besides being able to cause problems in the production, have the property of being bioaccumulative and being able to compromise the quality of insect meal and influence animal performance.

4. Practical developments: some examples on aquaculture nutrition

Many studies have been carried out to evaluate the use of insects in animal feed, including aquaculture. A study was performed with house fly larvae (*Musca domestica*) as a complimentary source of protein in Nile tilapia (*Oreochromis niloticus*) feeds. The authors observed superior growth rates (~3.76%/day) and reduced feed conversion ratio (1.05) possibly due to a better amino acid profile in this protein blend containing (28% fish meal, 25% house fly larvae, and 12% soy meal) [18]. In addition, the authors reported a high content of lipid (19.8%) in fly larvae, which should be considered when formulating diets. In a similar work, the replacement of 50–60% of fish meal by fly larvae meal (*Musca domestica*) in the feeding of tilapia fingerlings provided adequate growth and performance for the animals [20].

For feeding of African catfish (*Clarias gariepinus*), the larvae of flies have shown to be viable for their use [19]. However, the same positive response was not achieved by using butterfly larvae (*Bematistes macaria*) for feeding of African catfish, under experimental conditions [21]. The partial replacement of 40% fish meal with tenebrio meal for African catfish displayed no differences [22]. The animals have grown as well or better than those fed on the commercial diet. By partially replacing 25% of the fish meal with the tenebrio meal in gilthead seabream (*Sparus aurata*), no differences in weight gain and final weight were noticed [23]. However, for the 50% replacement level reduction in growth and specific growth rate, an increase in feed conversion was observed. One study tested levels of 25 and 50% of fish meal replacement by teneral flour for rainbow trout (*Oncorhynchus mykiss*) [34]. The results showed that there was no difference in performance and growth until the inclusion of 50% (isoprotein diets with 45% crude protein) [34].

For European juvenile sea bass (*Dicentrarchus labrax*), the inclusion of tenebrio meal at 25% had no adverse effects, but at 50% inclusion rate, the specific growth rate was reduced [35]. The use of tenebrio meal for Nile tilapia in partial replacement of fish meal at 25 and 50% levels reduced fish growth by around 29% [10]. According to the authors, the use of tenebrio meal for tilapia cannot be used in high proportions, because it is necessary to understand better the role of chitin in digestion and a better detection of possible toxins that can affect the growth of the fish. Another hypothesis may be related to the digestibility of tenebrio (FLT) meal, which in the form of the larvae's composition can influence its digestibility [36]. Studies evaluating the replacement of fish meal with giant tenebrio (*Zophobas morio*) for Nile tilapia obtained better feed conversion ratio and weight gain than the control, with ideal replacement value of 25%, which corresponded to 7.5% of inclusion of giant tenebrio meal in the feed [37].

Low survival rates were reported by other authors who worked with insect meal, for example, house fly larvae [38, 39] and tenebrio meals. The level of inclusion above 45% reduced survival to 70% [18]. The 50% fish meal replacement with FLT fish for common catfish (*Ameiurus melas* Raf.) resulted in inferior performance and survival, dropping of approximately 9% compared to the control (0% FLT) [40]. In contrast in rainbow trout, there was an increased survival with the inclusion of FLT but lower performance, digestibility, and alteration in the fillet fatty acid profile [24]. Jointly, these data suggest that in the early stages, FLT influences survival that is not pronounced in the final stages. In European juvenile sea bass (*Dicentrarchus labrax*), the inclusion of 25% FLT did not affect growth performance, while a higher inclusion level (50%) compromised the weight gain [25]; similar results were obtained in gilthead seabream juveniles, which included inclusions of 25–50%, compromising weight gain, specific growth rate, feed conversion efficiency, and protein efficiency ratio [23].

For tilapia [27], used the white larvae dry meal to formulate isonitrogenous and isoenergetic diets with maggot meal inclusions at 0, 30, 50, and 80 g/kg substituting gradually three conventional expensive feedstuffs: fish meal, fish oil, and soybean meal. The results showed no significant difference in growth parameters (final weight, weight gain, and SGR) and feed utilization efficiency (FCR and PER and feed intake) between treatments. Similarly fish whole body composition (dry matter, crude protein, lipid, ash, and fiber) was unaffected by the treatments except the fatty acid compositions which mirrored that of the diets. The cockroach (*Nauphoeta cinerea*) meal has also been tested for Nile tilapia with very promising results including superior zootechnical performance as compared to control diets [26].

Insect meals have also been evaluated in biofloc technology system [41–45]. As this system exhibits a series of particularities that separate it from the traditional clear and green water production systems such as recirculating aquaculture system, flow-throw, cages, and ponds, the following findings should be considered within the biofloc context and carefully extrapolated to other production systems. Levels higher than 10% of cockroach meal inclusion decrease the performance of the Nile tilapia juvenile in biofloc technology system, which may be related to the composition of the exoskeleton of the cockroach, especially chitin combined with sclerotin, which confers resistance and flexibility [41].

The use of tenebrio meal at 10% inclusion rate in the nursery stage of Nile tilapia in biofloc technology system did not affect the performance, somatic, hematological, and carcass composition indexes [42, 44]. Inclusion levels higher than 10% decreased productivity and survival and increased hepatosomatic index and lipid content, and in the carcass, consequence of the high lipid content and antinutritional factors is present in the tenebrio meal. Differently from the previous findings, a trial investigating gradual inclusion levels (0, 5, 10, 15, and 20%) of cockroach and tenebrio meal, individually, with *Litopenaeus vannamei* in biofloc technology system, concluded that juvenile shrimp accepted up to 15% of cockroach meal and up to 5% tenebrio meal [43, 45]. There is also a growing interest on the use of insects in shrimp feeds, as seen by several papers lately [46–50]. For additional and more scientific information, there are several papers on this topic [16, 51–54].

4.1 Constraints and future perspectives

Besides the cost and reliable commercial scale production, the diversity in terms of nutritional profile is considered one of the major issues of insect meal inclusion in aqua feeds. Some constraints were already discussed such as (i) excess lipid, (ii) amino acid imbalance, (iii) the presence of mycotoxins and possible antinutritional factors such as chitin [10], and the endogenous production of 1,4-benzoquinone toxin [55]. These isolated or combined factors may compromise the animal's immune system [10] and survival rates [56].

Chitin is an acetylated aminopolysaccharide similar to cellulose, but with a greater number of hydrogen bonds established with adjacent polymer chains, this confers extra resistance [56], which suggests greater difficulty in digestion. Tilapia fed diets with chitin and purified chitosan had impaired weight gain and feed conversion, and the chitin level of 2% was already harmful [57]. In addition, insects have between 11.6 and 137.2 mg kg^{-1} of body chitin [57]. The estimation done by [56] indicated an average value of 74.4 mg kg^{-1} of chitin in *Nauphoeta cinerea* meal, and when included in a ratio of 10%, inclusion will represent ~0.74% of dietary chitin, and for 20% inclusion will represent 1.50% of dietary chitin reaching critical levels. Some authors related high chitin levels with the reduction of feed consumption, availability of nutrients, and negative effects on performance [10, 22, 42, 44]. In this sense, it is fundamental to better understand the factors that limit the inclusion of insect meal into diets, either antinutritional or nutritional limit factors. Cost-effective formulations that met all animal requirements and selection of the other ingredients are crucial for good results. The diets isonitrogenous and isoenergetic with similar amino acids and fatty acids profile that met the nutritional requirement of the target species are fundamental points when comparing insect meal-based diets.

But one question remains unanswered: does insect in aquaculture nutrition a future eco-friendly approach or a commercial reality? The answer depends on the industry. For salmon, one of the biggest and high-value chain aquaculture sectors, the insect meal already offers an alternative to fish meal and soya in early stages of salmon production [46, 49, 50]. An example such as Skretting in Norway observed that fish showed the same zootechnical performance with feeds using insect meal as with traditional protein sources. The diets were normally made from the larvae of the black soldier fly. This feedstuff is an EU-approved commodity, and recent surveys show that Norwegian consumers are positive to eating salmon that has had insect meal in the feed formulation. In the future, an educational approach (e.g., focused on the blue economy) will play a major role to increase consumers' acceptance.

In order for such eco-friendly approach to reach another level, the industrial production of insect meal must be increased to meet the actual aquaculture demand. For example, in the European market, there is now little available insect meal for use on a large scale. To supply the feed mill demand, companies need to work together with manufacturers who wish to come up at a commercial level [46, 49, 50]. According to [46, 49, 50], by 2022 there will be at least five different European suppliers, each producing 20,000 tonnes of insect meal per year, that is, two thirds of the amount of soybean concentrate Skretting Norway uses today. In regard to other industries, although tests were successfully performed with tilapia [18, 27, 36, 37] and *L. vannamei* [46–50], the use of insects in a short-term due to massive volumes of feed demand for these industries is unlikely. The salmon example applying insects in early stages of production is one alternative that will be followed.

5. Conclusion

The use of insect meal in animal feed has been the subject of research, but the results are varied and divergent. Much is explained by the nutritional variability of insect meal production. In addition to knowing the nutritional values of insects, we must consider the study of insect nutrition, since depending on the species we can modulate the fatty acid profile with EPA and DHA and amino acid profile, mainly in the lysine, methionine, threonine, and tryptophan ratios. Understanding better the factors that limit the inclusion of insect meal, whether there is the presence of mycotoxins, fat rancidification, or unknown antinutritional factors, is one of the challenges to be understood. Some insects may have nutraceutical factors in their composition, which can confer exponential positive results on animal performance and human health. In this sense, the prospects of the use of insect meal in animal nutrition are very encouraging and new nutritional approaches are just starting.

Author details

André Freccia[1*], Jiovani Sergio Bee Tubin[2], Artur Nishioka Rombenso[3] and Maurício Gustavo Coelho Emerenciano[3,4]

1 Barriga Verde University Centre (UNIBAVE), Orleans, SC, Brazil

2 Animal Science Postgraduation Program, Santa Catarina State University (UDESC), Chapecó, SC, Brazil

3 Agriculture and Food, Aquaculture Program, Bribie Island Research Centre, Commonwealth Scientific and Industrial Research Organisation (CSIRO), Woorim, QLD, Australia

4 Nutrition Laboratory of Aquatic Organisms (LANOA), Santa Catarina State University (UDESC), Laguna, SC, Brazil

*Address all correspondence to: andrefreccia@gmail.com; paca@unibave.net

References

[1] Makkar HPS, Tran G, Heuzé V, Ankers P. State-of-the-art use of insects as animal feed. Animal Feed Science and Technology. 2014;**197**:1-33

[2] Khan S, Naz S, Sultan A, Alhidary IA, Abdelrahman MM, Khan RU, et al. Worm meal: A potential source of alternative protein in poultry feed. World's Poultry Science Journal. 2016;**72**(01):93-102

[3] Van Huis A, Van Itterbeeck J, Klunder H, Mertens E, Halloran A, Muir G, et al. Edible Insects: Future Prospects for Food and Feed Security (No. 171). Food and Agriculture Organization of the United Nations. Rome: FAO; 2013. ISSN: 0258-6150

[4] Yang Y, Yang J, Wu W, Zhao J, Song Y, Gao L, et al. Biodegradation and mineralization of polystyrene by plastic-eating mealworms: Part 1. Chemical and physical characterization and isotopic tests. Environmental Science and Technology. 2015;**49**(1):12080-12086

[5] Mora C, Tittensor DP, Adl S, Simpson AGB, Worm B. How many species are there on earth and in the ocean? PLoS Biology. 2011;**9**(8):e1001127. DOI: 10.1371/journal.pbio.1001127

[6] Ramos-Elorduy J. 2000 Current ethno-thermology in Mexico in human nutrition, traditional medicine and animal feed and recycling. In: Memories of the 35th National Congress of Entomology. pp. 3-46

[7] FAO. Insects for Food and Feed. 2015. Available from: <http://www.fao.org/forestry/edibleinsects/en/> (Accessed: 2 November 2015)

[8] Gatlin DM, Barrows FT, Brown P, Dabrowski K, Gaylord TG, Hardy RW, et al. Expanding the use of sustainable plant products in aquafeeds: A review. Aquaculture Research. 2007;**38**:551-579

[9] Gai F, Gasco L, Daprà F, Palmegiano GB, Sicuro B. Enzymatic and histological evaluations of gut and liver in rainbow trout, *Oncorhynchus mykiss*, fed with rice protein concentrate diets. Journal of the World Aquaculture Society. 2012;**43**:218-229

[10] MªJ S-M, Haro C, Sanz A, Braided CE, Villareces S, Barroso FG. Nutritional evaluation of Tenebrio molitor meal as fishmeal substitute for tilapia (*Oreochromis niloticus*) diet. Aquaculture Nutrition. 2016;**22**(5):943-955

[11] Hardy RW. Utilization of plant proteins in fish diets: Effects of global demand and supplies of fishmeal: Review article. Aquaculture Research, USA. 2010;**41**(5):770-776

[12] Veldkamp T, Van Duinkerken G, Van Huis A, Lakemond CMM, Ottevanger E, Bosch G, et al. Insects as a Sustainable Feed Ingredient in Pig and Poultry Diets - A Feasibility Study. Rapport 638–Wageningen Livestock Research; 2012. ISSN: 1570-8616

[13] Mancuso T, Baldi L, Gasco L. An empirical study on consumer acceptance of farmed fish fed on insect meals: The Italian case. Aquaculture International. 2016;**24**(5):1489-1507

[14] Diener S, Zurbrügg C, Tockner K. Conversion of organic material by black soldier f0ly larvae: Establishing optimal feeding rates. Waste Management & Research. 2009;**27**(6):603-610

[15] Sheppard DC, Newton GL, Thompson SA, Savage S. A value added management system using the black soldier fly. Bioresource Technology. 1994;**50**(3):275-279

[16] Henry M, Gasco L, Piccolo G, Fountoulaki E. Review on the use of

insects in the diet of farmed fish: Past and future. Animal Feed Science and Technology. 2015;**203**:1-22

[17] Collavo A, Glew RH, Huang YS, Chuang LT, Bosse R, Paoletti MG. House cricket small-scale farming. In: Paoletti MG, editor. Ecological Implications of Minilivestock: Potential of Insects, Rodents, Frogs and Snails. New Hampshire: New Hampshire Science Publishers; 2005. pp. 519-544

[18] Ogunji JO, Kloas W, Wirth M, Neumann N, Pietsch C. Effect of housefly maggot meal (magmeal) diets on the performance, concentration of plasma glucose, cortisol and blood characteristics of *Oreochromis niloticus* fingerlings. Journal of Animal Physiology and Animal Nutrition. 2008;**92**(4):511-518

[19] Aniebo AO, Erondu ES, Owen OJ. Replacement of fish meal with maggot meal in African catfish (*Clarias gariepinus*) diets. Revista Cientifica UDO Agricola. 2009;**9**(3):666-671

[20] Ezewudo BI, Monebi CO, Ugwumba AAA. Production and utilization of *Musca domestica* maggots in the diet of *Oreochromis niloticus* (Linnaeus, 1758) fingerlings. African Journal of Agricultural Research. 2015;**10**(23):2363-2371

[21] Nwamba H, Ogunji JO. Evaluating butterfly larvae (*Bematistes macaria*) meal as fishmeal substitute in diets of African catfish hybrid (Heteroclarias). Indian Journal of Social and Natural Sciences. 2012;**1**(1):78-84

[22] Ng WK, Liew FL, Ang LP, Wong KW. Potential of mealworm (*Tenebrio molitor*) as an alternative protein source in practical diets for African catfish, Clarias gariepinus. Aquaculture Research. 2001;**32**(1):273-280

[23] Piccolo G, Marono S, Gasco L, Iannaccone F, Bovera F, Nizza A. Use of Tenebrio Molitor Larvae Meal in Diets for Gilthead Sea Bream *Sparus aurata* Juveniles. The Netherlands: Insects to Feed The World; 2014. pp. 14-17

[24] Belforti M, Gai F, Lussiana C, Renna M, Malfatto V, Rotolo L, et al. Tenebrio molitor meal in rainbow trout (*Oncorhynchus mykiss*) diets: Effects on animal performance, nutrient digestibility and chemical composition of fillets. Italian Journal of Animal Science. 2015;**14**(4):4170

[25] Gasco L, Henry M, Piccolo G, Marono S, Gai F, Renna M, et al. Tenebrio molitor meal in diets for European sea bass (*Dicentrarchus labrax* L.) juveniles: Growth performance, whole body composition and in vivo apparent digestibility. Animal Feed Science and Technology. 2016;**220**:34-45

[26] Freccia A, Meurer ES, Jerônimo GT, Emerenciano MGC. Insect meal in diets of tilapia fingerlings. Zootechnics Archives. 2016;**65**(252):541-547

[27] FAO (Food and Agriculture Organization of United Nations). Edible insects: Future prospects for food and feed security. Forestry Paper. 2013;**171**:1-154

[28] Potrich TD, Lorini I, Voss M, Steffens MCS, Pavani DP. Methodology for the Creation of *Tenebrio molitor* in Laboratory to Obtain Larvae. Embrapa - Wheat, Online Documents no 82. Dec, 2007. Available at <http://www.cnpt.embrapa.br/biblio/do/p_do82_12.htm> (Accessed: 19 October 2016)

[29] Nutrinsecta. Creation of Tenebrio Molitor. 2013. Available from: <http://www.nutrinsecta.com.br/artigos/criacao-do-tenebrio-molitor/> (Accessed: 19 October 2016)

[30] Urs KCD, Hopkins TL. Effective of moisture on growth rate and development of two strains of Tenebrio molitor L. (Coleoptera, Tenebrionidae).

Journal of Stored Products Research. 1973;**8**(4):299-305

[31] Ramos-Elorduy J, Avila Gonzalez E, Rocha Hernandez A, Pino JM. Use of Tenebrio molitor (Coleoptera: Tenebrionidae) to recycle organic wastes and feed for broiler chickens. Journal of Economic Entomology. 2002;**95**(1):214-220

[32] Barroso FG, de Haro C, Sánchez-Muros MJ, Venegas E, Martínez-Sánchez A, Pérez-Bañón C. The potential of various insect species to use as food for fish. Aquaculture. 2014;**422**:193-201

[33] Furuya WM. Brazilian tables for tilapia nutrition. GFM. Toledo. 2010;**100**:63-82

[34] Gasco L, Belforti M, Rotolo L, Lussiana C, Parisi G, Terova G, et al. Mealworm (Tenebrio molitor) as a potential ingredient in practical diets for rainbow trout (*Oncorhynchus mykiss*). In: Insects to Feed the World, The Netherlands, 14-17 May 2014. pp. 78; 2014

[35] Gasco L, Gai F, Piccolo G, Rotolo L, Lussiana C, Molla P, et al. Substitution of fish meal by Tenebrio molitor meal in the diet of *Dicentrarchus labrax* juveniles. In: Insects to Feed the World, The Netherlands, 14-17 May 2014. p. 80; 2014

[36] Adesulu EA, Mustapha AK. Use of housefly maggots as a fishmeal replacer in tilapia culture: A recent vogue in Nigeria. In: 5th International Symposium on Tilapia Aquaculture, Rio de Janeiro, Brazil, 1, 138; 2000

[37] Jabir MDAR, Razak SA, Vikineswary S. Nutritive potential and utilization of super worm (*Zophobas morio*) meal in the diet of Nile tilapia (*Oreochromis niloticus*) juvenile. African Journal of Biotechnology. 2012;**11**(24):6592-6598

[38] Idowu AB, Amusan AAS, Oyediran AG. The response of Clarias gariepinus fingerlings (Burchell 1822) to the diet containing housefly maggot (*Musca domestica* L). Nigerian Journal of Animal Production. 2003;**30**(1):139-144

[39] Ajani EK, Nwanna LC, Musa BO. Replacement of fish meal with maggot meal in the diets of Nile tilapia, Oreochromis niloticus. World Aquaculture. 2004;**35**:52-54

[40] Roncarati A, Gasco L, Parisi G, Terova G. Growth performance of common catfish (*Ameiurus melas* Raf.) fingerlings fed mealworm (*Tenebrio molitor*) diet. Journal of Insects as Food and Feed. 2015;**1**(3):233-240

[41] Tubin JB, Gutierrez SM, Monroy-DostaMC, EmerencianoMGC. Cockroach (*Nauphoeta cinereal*) insect meal in diets for Nile tilapia raised in biofloc technology system. Aquaculture and Fisheries. 2019 (in review)

[42] JSB T, Paiano D, GSO H, Furtado WE, Martins ML, Durigon E, et al. *Tenebrio molitor* meal in diets for Nile tilapia juveniles reared in biofloc system. Aquaculture. 2019:734763

[43] Molinari D, Sgnaulin T, de Mello GL, Emerenciano MGC. Specked cockroach Nauphoeta cinereal and mealworm Tenebrio molitor meals in diets for Pacific white shrimp Litopenaeus vannamei reared in biofloc technology system. Aquaculture International. 2019 (in review)

[44] Kroeckel S, AGE H, Roth I, Katz H, Wuertz S, Susenbeth A, et al. When a turbot catches a fly: Evaluation of a pre-pupae meal of the black soldier fly (*Hermetia illucens*) the fish meal substitute- growth performance and chitin degradation in juvenile turbot (Psetta maxima). Aquaculture. 2012;**364/365**:345-352

[45] Devic E, Leschen W, Murray FJ, Little DC. Growth performance, feed utilization and body composition of advanced nursing Nile tilapia (*Oreochromis niloticus*) fed diets containing black soldier Fly (*Hermetia illucens*) larvae meal. Aquaculture Nutrition. 2018;**24**:416-423. DOI: 10.1111/anu.12573

[46] Cummins VC Jr, Rawles SD, Thompson KR, Velasques A, Kobayashi Y, Hager J, et al. Evaluation of black soldier fly (*Hermetia illucens*) larvae meal as partial or total replacement of marine fish meal in practical diets for Pacific white shrimp (*Litopenaeus vannamei*). Aquaculture. 2017;**473**:337-344

[47] Panini RL, Freitas LEL, Guimaraes AM, Rios C, da Silva MFO, Vieira FN, et al. Potential use of mealworms as an alternative protein source for Pacific white shrimp: Digestibility and performance. Aquaculture. 2017;**473**:115-120

[48] Panini RL, Pinto SS, Nobrega RO, Vieira FN, Fracalossi DM, Samuels RI, et al. Effects of dietary replacement of fishmeal by mealworm meal on muscle quality of farmed shrimp *Litopenaeus vannamei*. Food Research International. 2017;**102**:445-450

[49] Finke MD. Estimate of chitin in raw whole insects. Zoo Biology. 2007;**26**(2):105-115

[50] Skretting Website: Insect Meal Now in Commercial Skretting Feed. Available from: https://www.skretting.com/en/settings/news/general-news/insect-meal-now-in-commercial-skretting-feed/

[51] Arru B, Furesi R, Gasco L, Madau FA, Pulina P. The introduction of insect meal into fish diet: The first economic analysis on European Sea bass farming. Sustainability. 2019;**11**:1697

[52] Nogales-Merida S, Gobbi P, Jozefiak D, Mazurkiewicz J, Dudek K, Rawski M, et al. Insect meal in fish nutrition. Reviews in Aquaculture. 2018:1-24

[53] Pinotti L, Giromini C, Ottoboni M, Tretola M, Marchis D. Insects and former foodstuffs for upgrading food waste biomasses/streams to feed ingredients for farm animals. Animal. 2019;**13**(7):1365-1375

[54] Tran G, Heuze V, Makkar HPS. Insects in fish diets. Animal Frontiers. 2015;**5**(2):37-44

[55] El-Kashlan IH, Helal E, El-Hamid A, Abdellatif MA. Teratogenicity and embryotoxicity of 1,4-benzoquinone and wheat flour infested with Tenebrionid beetles on mice. Journal of King Saud University. 1996;**8**:121-133

[56] Barker D, Fitzpatrick MP, Dierenfeld ES. Nutrient composition of selected whole invertebrates. Zoo Biology. 1998;**17**(2):123-134

[57] Shiau S, Yu Y. Dietary supplementation of chitin and chitosan depresses growth in tilapia, *Oreochromis niloticus* × *O. aureus*. Aquaculture. 1999;**179**(1):439-446

Chapter 10

Tackling of Illegal Fishing in Shallow Waters

Arun Raj Velraj

Abstract

Illegal fishing is one of the most prevailing problems across the oceans of the world. It causes both material loss and the loss of marine richness. There are many organizations that are concerned about this issue all around the world. All these organizations go for very complicated and costly solutions, such as satellite monitoring and image processing of the ships that involved in the illegal fishing activities all over the world. Here, we have gone for a simple as well as an effective solution in which we will detect the operation of the motors by attaching a live sensor to it, thereby continues monitoring of the boat becomes possible for the officials to monitor the boats throughout the day without involving complicated approaches that are costlier than this approach. This project not only involves monitoring systems, whereas there is also a provision to prevent the boat from involving in illegal activities. I have used for wireless speed control of the motor which is a very approach to prevent the boats from involving in illegal fishing. This can save the marine animals from extension or endangerment by many folds in total.

Keywords: remote monitoring, unregulated angling, speed control and alert systems

1. Introduction

The fishery is a run of the mill case [1] of a typical property asset, and along these lines experiences the normal property issue. Individual anglers have motivations to gather more than what is socially ideal, since they do not consider the stock externality of reaping. To manage overfishing [2] and overcapacity, the administration experts have presented guidelines, including rigging and exertion limitations, territory confinements [3], landing charges, reap quantities, least sizes, by-get guidelines, and so on.

Fishery guidelines [4] are not, by and large, self-upholding. Neither other anglers nor the group of other oceanic vessels educate the specialists of encroachments they witness. This implies implementation costs must be brought about. As the expense of authorization increments with the dimension of prevention, it probably will not be ideal to have full consistence as the target. This is talked about in more prominent detail in consequent segments. The issue of rebelliousness with guidelines emerges on the grounds that the presentation of principles and guidelines in the fishery does not consequently imply that the specialists' motivating forces to disregard guidelines are evacuated, despite the fact that the anglers' motivators absolutely can be changed.

With global warming being the center of the marine ecosystem, in addition to this, illegal fishing also adds to the marine ecosystem's destruction on a whole. The type of illegal fishing [5] that is being addressed in this project involves fishing the oceans during the breeding seasons of the year. This has many adverse effects on the marine as well as the earth's ecosystem as a whole. This act reduces the amount of fish that is being regulated inside the oceans [6] as a whole leading to a collapse in the prey predator ratio, which is a very important aspect since the disturbance in this ratio can adversely affect the marine ecosystem [7] as well as the earth's ecosystem as a whole. This phenomenon is clearly being monitored and prevented in three distinct and independent modules operating as a single device.

2. Proposed method

The approach we used to detect the movement of the ships was to fix a vibrational sensor to the motor that is present in the boats that go into the ocean to fish. We went for fixing the vibrational sensors in the boat's motor because it is one of the most common things that is present in all the boats. As far as the Indian fishing boats are concerned, 99% of these boats are propelled by motors that are powered by a diesel engine [8]. The motors are prone to produce vibrations of their own in addition to the vibrations [9] that are produced by the boats when it is operational. But in the current scenario, boats comes with pre-determined vibrational arresters and hence we come with a novel approach of tackling illegal angling through the block diagram shown in **Figure 1**.

As mentioned in **Figure 1**, the proposed approach is categorized into three modules:

1. First module involves with detection of movement of boats by interfacing the ADXL345 (Accelerometer sensor) with the Arduino Uno board. This module gives the analog value of sensor when the boat moves in the water.

2. Second module involves with the updating the analog values into the RPi Server through node MCU (Wi-Fi module) in the form of tables. The values are updated with the time at which the value is detected in the database. These values are compared with the pre-defined values. Here, the pre-defined values include the prohibited times that is approved by government to prevent the fisherman from fishing during breeding times in order to protect the marine ecosystem.

3. Third module involves with the speed control [10] of the motor. This is achieved by controlling the motor wirelessly by sending the signal to Arduino Uno through node MCU (Wi-Fi module). The speed of the motor is varied by changing the

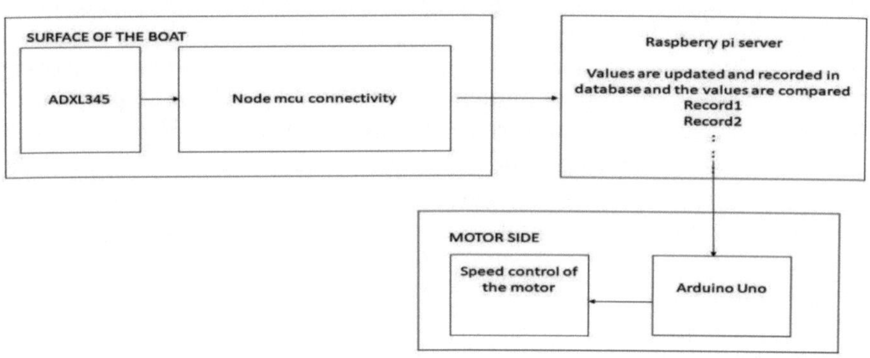

Figure 1.
Overall model of the proposed work.

firing angle. By increasing the firing angle, the speed of the motor can be increased; and by decreasing the firing angle, the speed of the motor is decreased.

3. Results and discussions

The proposed method is preceded with the above-mentioned working process and the findings are discussed here.

The entire setup of the project that is involved in controlling the speed of the motor is shown in **Figure 2**. The sensor AD XL345 is interfaced with Arduino and it is used for the calculation of movement of boats. The output is generated by the movement of ship. The output is generated when the ADX1345 is moved, and results are displayed in serial monitor, as shown in **Figure 3**.

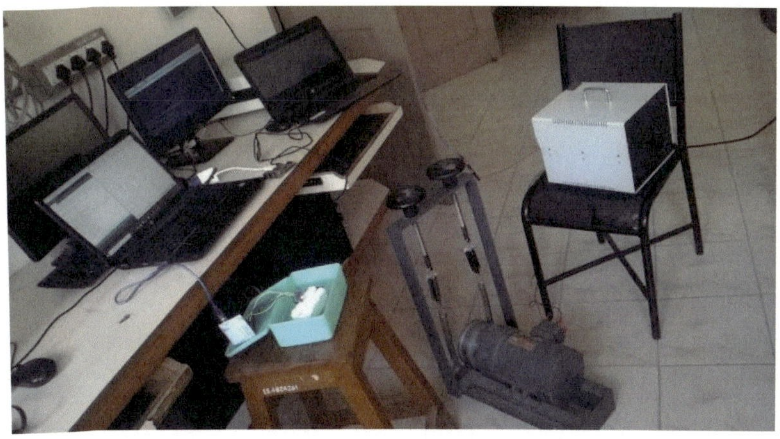

Figure 2.
Setup of the project.

COM25 (Arduino/Genuino Uno)

x = 320	y = 321	z = 321	Roll = 19.47	Pitch = 69.33
x = 320	y = 321	z = 320	Roll = 19.10	Pitch = 69.72
x = 318	y = 320	z = 319	Roll = 19.77	Pitch = 68.08
x = 318	y = 320	z = 319	Roll = 19.77	Pitch = 68.08
x = 318	y = 320	z = 319	Roll = 19.77	Pitch = 68.08
x = 319	y = 320	z = 320	Roll = 20.14	Pitch = 68.69
x = 319	y = 320	z = 320	Roll = 20.14	Pitch = 68.69
x = 318	y = 320	z = 319	Roll = 19.77	Pitch = 68.08
x = 318	y = 319	z = 318	Roll = 20.41	Pitch = 68.47
x = 318	y = 319	z = 319	Roll = 20.78	Pitch = 68.08
x = 74	y = 18	z = 0	Roll = 40.88	Pitch = 54.34
x = 234	y = 347	z = 283	Roll = 354.04	Pitch = 39.42
x = 297	y = 319	z = 309	Roll = 17.52	Pitch = 55.48
x = 315	y = 321	z = 318	Roll = 18.39	Pitch = 65.59
x = 320	y = 323	z = 321	Roll = 17.31	Pitch = 69.33
x = 321	y = 323	z = 322	Roll = 17.66	Pitch = 70.00
x = 318	y = 320	z = 319	Roll = 19.77	Pitch = 68.08

Figure 3.
Output of ADXL345.

Firing angle (%)	Speed of motor (RPM)	Status of motor
0.4	10	OFF
25	400	ON
45	700	ON
75	1000	ON

Table 1.
Variation of motor speed based on firing angle.

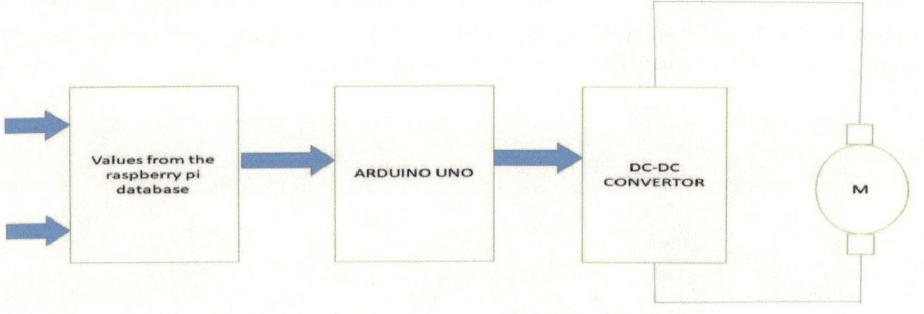

Figure 4.
Speed control of a motor.

The values from the ADXL345 are updated through the raspberry pi through MYSQL language. The values are stored in the form of tables with the ANALOG value of sensor and the time at which the boat enter into the sea. The obtained results are compared with the pre-defined data stored in the database and a trigger pulse is given to the motor to control the firing angle as shown in **Table 1**.

The speed of the motor is controlled by the PWM technique [11, 12]. This is done by varying the firing angle. The beacon signal produced is of 5 V which is not sufficient to control the 115 V motor and thus the signal is given to the DC-DC, which boosts the produced signal to control the 115 V DC motor, as described in **Figure 4**.

The motor is varied wirelessly to tackle illegal fishing activity. The motor speed is varied by adjusting the firing angle of the PWM signal. This will then freeze the motor of the boat and prevent the fisherman in entering into the sea.

4. Conclusion and future scope

This approach helps in the prevention and tackling of illegal fishing activities of fishermen in the water surfaces. The first phase of the project involves in the prevention of the illegal fishing, and the other phase involve in the tackling of illegal fishing. The proposed idea is very efficient and it is applicable for every boat in fishing. This is very cheap and cost effective. Another main advantage of the project is that the boats are prevented from involving in the illegal activities at a very close distance of 50–60 m from the coastal region. This will reduce the risk of illegal fishing as well as the boats will be easily detected. The stopped boats can also be retrieved from the ocean because the boats have been topped from the ocean at an accessible distance from the seashore. This approach can be further extended by attaching a solar panel to the module to replace the batteries and tampering mechanisms can be included.

Author details

Arun Raj Velraj
Department of ECE, Mepco Schlenk Engineering College, Sivakasi, Tamil Nadu, India

*Address all correspondence to: arunraj@mepcoeng.ac.in

References

[1] Petrossian GA. Preventing illegal, unreported and unregulated (IUU) fishing: A situational approach. Biological Conservation. 2015;**189**:39-48

[2] Daud P et al. Concept development of integrated monitoring system model to support activities monitoring in the Border Region. In: 2015 International Conference on Smart Sensors and Application (ICSSA). IEEE; 2015

[3] Agnew DJ et al. Estimating the worldwide extent of illegal fishing. PLoS One. 2009;**4**(2):e4570

[4] Bacharuddin F, Wuryanto H, Nugraha B. Optimum work frequency for marine monitoring based on genetic algorithm. Telkomnika. 2018;**16**(4):1551-1559

[5] Akinbulire T et al. Responding to illegal, unreported and unregulated fishing with evolutionary multi-objective optimization. In: 2018 IEEE International Conference on Computational Intelligence and Virtual Environments for Measurement Systems and Applications (CIVEMSA); 2018

[6] Marzuki MI et al. Fishing gear recognition from VMS data to identify illegal fishing activities in Indonesia. In: OCEANS 2015—Genova. IEEE; 2015

[7] Marzuki MI et al. Fishing gear identification from vessel-monitoring-system based fishing vessel trajectories. IEEE Journal of Oceanic Engineering. 2018;**43**(3):689-699

[8] Arun TA, Madhukar TS, Deka J. Speed control of DC motor using PIC 16F877A microcontroller. Multidisciplinary Journal of Research in Engineering and Technology. 2014;**1**(2):223-234

[9] Liu Z, Jiang L. PWM speed control system of DC motor based on AT89S51. In: Proceedings of 2011 International Conference on Electronic & Mechanical Engineering and Information Technology. Vol. 3. IEEE; 2011

[10] Guangxian C. A design PWM speed controller of DC motor using single-chip microcontroller. Journal of Zhanjiang Normal College. 2008;**6**:123-125

[11] Petru L, Mazen G. PWM control of a DC motor used to drive a conveyor belt. Procedia Engineering. 2015;**100**:299-304

[12] Haishui Z et al. Design on a DC motor speed control. In: 2010 International Conference on Intelligent Computation Technology and Automation. Vol. 2. IEEE; 2010. p. 36